Digital Functions and Data Reconstruction

Li M. Chen

Digital Functions and Data Reconstruction

Digital-Discrete Methods

 Springer

Li M. Chen
University of the District of Columbia
Washington, DC, USA

ISBN 978-1-4899-9495-0   ISBN 978-1-4614-5638-4 (eBook)
DOI 10.1007/978-1-4614-5638-4
Springer New York Heidelberg Dordrecht London

Printed on acid-free paper

Springer is part of Springer Science+Business Media (www.springer.com)

To my mentors and colleagues who encouraged me to continue the research in this specific area.

To my mentors and colleagues who encouraged me to continue the research in this specific area.

Preface

Before Newton-Leibnitzs time, mathematics was basically "discrete." Since then, continuous mathematics has dominated the literature. But discrete mathematics has found new life with the appearance and widespread use of the digital computer. However, we still perfer to use the thinking involved in continuous mathematics. For example, if we had discrete information on some samples, we would assume a continuous model to do the calculation. Sometimes, we need a discrete output from the continuous solution, and it is not hard to re-digitize the continuous results. For some problems, going from "discrete" to "continuous" back to "discrete" may not always be necessary. In such instances, we can directly employ a methodology to go from "discrete input" to "discrete output."

The tractability and practice of the methodology using such a philosophy is certainly valid. Lets consider an example. In seismic data processing, the seismic data sets consist of synchronous records of reflected seismic signals registered by a large number of geophones (seismic sensors) placed along a straight line or in the nodes of a rectangular lattice on the earths surface. A series of explosions serve as the source of the initial seismic pulse, responses to which are averaged in a special manner. The vertical time axis forming the resulting two- or three-dimensional picture is identified with depth, so that the peculiarities of the reflected signal under the respective sensor carries information on the local properties of the rock mass at the respective point of the underground medium. In contrast to the above-lying sedimentary cover, the absence of pronounced reflecting surfaces in a crystalline body makes it difficult to infer the geological information from the basement interval of the seismic picture. We can see that layer description (or modeling) becomes a central problem. If we know a target layer in each horizontal or vertical (line) profile, we can get the entire layer in the 3D stratum. It can be transferred into a surface fitting problem where we can use a Coones surface, Bezier polynomial, or B-spline to fit the surfaces.

Based on the boundary values to fill the interior, the most suitable technique is a Coones surface. However, for a layer, one must make two surfaces, one for the top of the layer and one for the bottom. The Coons surfaces have no property of preserving a fitted surface in the convex of a guiding point set. That is to say, the upper surface may intersect with the lower surface. That is not a desired solution. Since there are many sampled points on measured lines, the Bezier polynomial is also not a good choice. One cannot make the order of the polynomial very high. B-spline is a very good choice for the problem, but we need to do a pre-partition and coordinate transformations. In fact, for the problem, we have no special requirements for the smoothness, and we just need two reasonable surfaces to cover the layer. Another example is from computer vision. In observing an image, if you extract an object from the image, a representation of the object can sometimes be described by its boundary curve. If all values on the boundary are the same, then we can just fill the region. If the values on the boundary are not the same, and if we assume that the values are "continuous" on the boundary, then one needs a fitting algorithm to find a surface. How do we fill it? Its solution will directly relates to a famous mathematical problem called the Dirchet's Problem and have direct application to in data compression.

If the boundary is irregular, the 2D B-spline needs to partition the boundary into four segments to form a XY-plane vs UV-plane translation. The different partitions may yield different results. Practically, the procedure of a computation is a set of "discrete" actions. The input of a curve is also discrete, and the output is discrete. We can, therefore, make the following arguments. Do we always need a continuous technique for surface fitting? Is it possible to have a discrete fitting algorithm to get a reasonable surface for the above problems? In 1989, L. Chen developed an algorithm to do such discrete surface fitting in 2D. The algorithm is called gradually varied fitting.

Gradually varied fitting was based on so called gradually varied functions that is a type of digital functions in general sense. In 1986, A. Rosenfeld invented a basic type of the digital continuous function for the purpose of image segmentation where one is to find a continuous-looking part in a digital image, a digital space.

This is book is written to different interest groups of readers. Chapters 1–3 are foundations for the entire book; Chaps. 4 and 5 are for senior students, graduate students, or researchers who are interested in digital geometry and topology. Chapter 6 is a knowledge foundation for data reconstruction. Chapters 7 and 8 is for senior students in scientific computing. Chapter 9 will not be difficult for graduate students in computer science or senior students in mathematics with computer graphics background. Chapter 10 is for senior students in mathematics. Chapters 11 and 12 deal with future topics. For the Chapters marked with "*" may need some advanced knowledge.

Chapters 1–3, 6 and 7 are basic chapters. Chapters 4 and 5 are in discrete mathematics especially discrete geometry and topology. Chapters 8–12 are application related topics in scientific computing.

Acknowledgments Many Thanks to my friends Liang Chen and Xiaoxi Tang, they helped and observed the publishing of my very first article related. That time was a difficult time to me in changing my work from a mathematical department back to computer science department.

Many thanks to my daughter Boxi Cera Chen who helped me to check English for the whole book. After her college, her professional editing tone has changed a bit to literal. That gives me some hard time to change back to simple sentences since most of our math and cs readers love simple sentences! We might already have a hard time to understand tough mathematics in this book. Nevertheless, she is the best I can found; she has both music and statistics degrees.

Many thanks to my wife, she has always supported my research, especially in science and math. Ten years also I was finishing my first English book "Discrete Surfaces and Manifolds." I have put special thanks to my boy Kyle Landon Chen, he was just born at that time. So I found some excuses to stay home a few months no come to UDC, then I could finish bit more of that book. Now today, when I ask Kyle to help dad for some sentences, he always rejected. Why me? Ask Cera!

One day I suddenly understanded! I bosted my understanding level of philosophy on human. If there is a way to escape, they will rather do it. If there is an old way of doing, people is reluctant to use a new method. That give me an idea to promote the theory of digital functions. We need to find more important real world problems in which digital techniques are the best or unique!

Meanwhile, we also need to make things available when people is looking for it. That is a purpose of this book!

Washington, DC, USA Li M. Chen

Acknowledgements

The author would like to thank many colleagues and researchers for their support. Special thanks to my friend Professor David Mount at University of Maryland, he has been supported my research for more than 15 years. He also reviewed some of the main results in the related research, which will be presented in Chaps. 3 and 4. As a Professor A. Rosenfeld's close associates, he knows the best of the methodology of the founder of digital functions.

Contents

Acronyms

N	The natural number set
I	The integer number set
R	The real number set
$G = (V, E)$	A graph G with the vertex set V and the edge set E
D	A simply connected domain
J	A subset of D, J usually indicates the sample points or the guiding points
GVS	Gradually varied surfaces
GVF	Gradually varied functions
Σ_m	m-dimensional digital space

Part I
Digital Functions

Chapter 1
Introduction

Abstract Digital continuous functions and gradually varied functions were developed in the late 1980s. A. Rosenfeld [24] proposed digital continuous functions for digital image analysis, especially to describe the "continuous" component of a digital image, which usually indicates an object. L. Chen [6] invented gradually varied functions to interpolate a digital surface when sample points in its boundary appear to be gradually varied. In this introduction chapter, we will describe the necessity of developing such a method and its relationship to modern numerical analysis and even functional analysis. We will also discuss the various applications of developing this theory and its role in predicting future trends.

1.1 Overview

This book provides a solid foundation for the theory of digital functions and its applications to image data analysis, digital object deformation, and data reconstruction. Digital continuous functions and gradually varied functions were developed in the late 1980s [6, 24]. A. Rosenfeld [24] proposed digital continuous functions for digital image analysis, especially to describe the "continuous" component of a digital image, which usually indicates an object [24]. Independently, L. Chen (1989) invented gradually varied functions to interpolate a digital surface when sample points in its boundary appear to be gradually varied [6].

This new method has a unique feature: it is mainly built on discrete mathematics with connections to classical methods in mathematics and the computer sciences. During the past 25 years, these concepts have undergone experienced significant development in both theory and practice. There are three major milestone developments:

(1) Defining digital continuous functions and gradually varied functions. In 1986, A. Rosenfeld proposed digital continuous functions for digital image analysis [24]. L. Chen (1989) invented gradually varied functions to interpolate a digital surface [6]. More importantly, the necessary and sufficient conditions for

the existence of gradually varied extension were discovered in 1989. Recently, G. Agnarsson and Chen found the relationship between gradually varied functions and graph homomorphism, in addition to proving the overall extension theorems [1].

(2) E. Khalimsky, T. Y. Kong, L. Boxer, Rosenfeld et al. developed methods for digital deformations, especially for digital homotopy including the calculation of topological groups [4, 21, 22, 25]. Many papers have been published; this area has great potential not only in practical image processing but also in the development of algorithms for homotopy group calculations of real data. Calculation of homotopy groups is a world-class unsolved problem. Homotopy groups generally pose challenging problems, however they may be less problematic in digital spaces. Even though a unified theory in this area has not yet been discovered, it is very likely researchers will soon make a breakthrough that will contribute to the entire field of Mathematics.

(3) In 2010, Chen designed a systematic digital-discrete method for obtaining continuous functions with smoothness of a certain order ($C^{(n)}$) from sample data [13]. This method is based on gradually varied functions and the classical finite difference method. This new method has been applied to real groundwater data and the results have validated the method. This method is also independent from existing popular methods such as the cubic spline method and the finite element method. The new digital-discrete method has considerable advantages in a large number of real data applications. Chen and his colleagues have written a series of research papers from regular rectangular domains to arbitrary manifolds, from continuous functions to harmonic functions, and from applications in ground water data to current applications in wave equations. Specifically, Chen and Luo [16] proposed using harmonic functions to extend the role of piecewise linear functions (such as triangulation) in reconstructing functions on manifolds [16].

In theory, gradually varied functions and digital continuous functions are very similar. Gradually varied functions are more general in terms of being functions of real numbers, while digitally continuous functions are easily to map one digital space to another.

This book gives an introduction and comprehensive coverage of digital function methods. We also provide scientists and engineers who deal with digital data a highly accessible, practical, and mathematically sound introduction to the powerful theories of digital topology and functional analysis, while avoiding the more abstruse aspects of these topics.

1.2 Why Digital Continuity and Gradual Variation

A good theory is usually developed based on a simple need to solve a given problem; its resolution often brings a general principle that solves a large number of real world problems.

The basic task in digital image processing is to find an object. An object usually a specific coloring or pattern that separates it from surrounding background or other objects. Since the image is digitally stored in computer memory or disks, it is essential to describe the continuity of a digitally represented object.

The question becomes, why are continuous functions unable to deal with such a digital object directly? The reason is very simple, there is an infinite number of continuous functions that lie on those digital points (samples). To find a best or near best continuous function that goes through these digital sample points is time consuming.

On the other hand, if we had discrete information on some samples, we would assume a continuous model to do the calculation. When we need a discrete output from the continuous solution, we will re-digitize the continuous results. There are many examples of this in computer graphics models and displays, the most popular of which is spline fitting.

For some problems, going from "discrete" to "continuous" and then back to "discrete" may not always be necessary. For other problems, the continuous model is not obvious and may be very difficult to find. This question always brings up problems in scientific computing.

A simple example is to reconstruct a volume from a geophysical measurement data, see Fig. 1.1 [11]. Figure 1.1a shows that the layer described by perpendicular lines, so we can apply a direct fitting; Fig. 1.1b shows that the layer described by arbitrary lines, where a re-sampling or domain transformation needs to be applied.

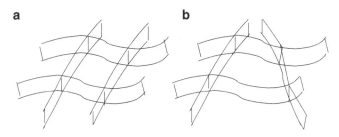

Fig. 1.1 (**a**) The layer described by perpendicular lines, one can apply a direct fitting; (**b**) the layer described by arbitrary lines, one must apply a resampling or domain transformation

Why do we not directly employ a methodology to go from "discrete input" to "discrete output"? The tractability and practice of the methodology using such a philosophy is certainly valid.

The direct treatment of digital objects using digital functions is necessary. Digital continuous functions were first proposed by Rosenfeld in 1986 [24]. This type of function was first seen in digital images to describe smooth-looking part which appears to be a "continuous" function in digital space. Based on Rosenfeld's concept, L. Boxer in 1994 [4] had developed digitally continuous functions for topological movement [2] of digital curves.

On the other hand, gradually varied functions were defined by Chen in citeAle of digital curve. It was a simplification based on the so called λ-connected set defined by Chen in 1985 [9, 11]. For example, in order to find a velocity layer in real seismic wave data, the velocity of the wave can be changed gradually in a big layer. λ-connectedness is to define a relatively smooth change from one point to another through a path. This technique is called λ-connected segmentation. Using the velocity layer information, parameters were extracted inside a layer to determine rock types and oil-gas probabilities. A new problem then surfaced: how to reconstruct a 3D data volume to indicate the layer location information. Not only do we need to interpolate digital surfaces in order to do this, but we also need to re-build the volume. In 1989, Chen found a necessary sufficient condition to interpolate a discrete surface without using continuous space assumptions [6]. This is the gradually varied surface interpolation where the gradually varied surface is a kind of λ-connected set [9].

To summarize, from image analysis needs, Rosenfeld proposed digitally continuous functions. But in order to reconstruct an image, Chen proposed gradually varied functions.

1.3 Basic Concepts of Digital Functions

The concept of the "continuous" function in digital space (also known as digitally continuous functions or digital continuous functions) was proposed by A. Rosenfeld in 1986 [24]. It describes a function whose values are integers on digital points. The values are the same or about the same as those of its neighbors. In other words, if x and y are two adjacent points in a digital space, then $|f(x) - f(y)| \leq 1$. If the function is a mapping from a connected set to another, then $d(x,y) \leq 1$ implies $d(f(x), f(y)) \leq 1$.

A gradually varied function is more general than a digital continuous function. A gradually varied function is a function from a digital space Σ to $\{A_1, A_2, \cdots, A_m\}$ where $A_1 < A_2 < \cdots < A_m$ and A_i are rational or real numbers [6]. This function possesses the following property: If x and y are two adjacent points in Σ, assume $f(x) = A_i$, then $f(y) = A_i$, A_{i+1}, or A_{i-1}.

This generalization plays an important role in smooth data reconstruction. In its methodology, a discrete method will be added to a digital data fitting method, it will be called digital-discrete data reconstruction. A digital method is semi-symbolic and semi-numeric since the digital method uses the levels of values and graph-theoretical methods. The levels of values can sometimes be viewed as symbols; also have numerical meaning not only in terms of order but also in terms of possessing actual, real number values. The disadvantage of a digital method is that it can not easily perform an approximation when we want to use derivatives. That is why we want to add a discrete method on top of the digital method [12–14].

An extension theorem related to the above functions was completed by Chen in 1989 [6–8]. This theorem states: Let $J \subset \Sigma$ and $f : J \rightarrow \{A_1, A_2, \cdots, A_m\}$. The necessary and sufficient condition for the existence of the gradually varied extension

F of f is: for each pair of points x and y in J, assume $f(x) = A_i$ and $f(y) = A_j$, we have $|i - j| \leq d(x,y)$, where $d(x,y)$ is the (digital) distance between x and y.

It is believed that Rosenfeld had attempted to get a result that related to this important theorem [24]. The gradually varied surface has a direct relationship to graph homomorphism [1].

1.4 Gradually Varied Functions and Extensions

There are many important properties in gradually varied functions. Some theorems are mathematically interesting. The envelope theorem, a uniqueness theorem, and an extension theorem that concerns preserving the same norm, are shown in this book. Gradually varied functions can be induced from constructive continuous functions in constructive mathematics [3, 11].

In this book, we also introduce gradually varied extensions or normal immersions. This studies the mapping between two discrete objects, especially discrete manifolds.

Two key theorems related to this topic are presented: (1) Any graph can normally immerse into an arbitrary tree T. (2) Any graph can normally immerse into the grid space with indirect adjacency.

These theorems have potential applications in data reconstruction. In practice, we will show an optimal uniform approximation theorem of gradually varied functions and develop an efficient algorithm for the approximation.

Since the book mainly focuses on basic methods and data reconstruction, we will give brief descriptions of the profound aspects of gradually varied functions and extensions. Recently, the more general cases of the extension were studied with relation to graph homomorphism [1].

1.5 Digital Curve Deformation

Digital curve deformation is related to the "continuous" changes of a digital curve. This concept is essential to image processing and computer graphics. For instance, image morphing is a common technology for digital movie production. Even though a precise definition for digital deformation is still an open problem, a great deal of research accomplishments have already been achieved.

Khalimsky [21] first studied the digital curve deformation for defining digital fundamental groups [21]. Kong has a more general approach in this research [22]. Boxer [4] used Rosenfeld's digital continuous function to define digital deformation and digital homotopy [4]. His definition was to simulate the mathematical definition for homotopy in continuous space. For 2D images, in 1996 Rosenfeld first defined that local deformation for two digital curves C and D as every pixel (point) of C coincides with or is a neighbor of a pixel (point) of D, and vice versa [25]. This is an intuitive definition; however, it does not work in every case. It is close to a necessary condition. Rosenfeld and Nakamura [22] modified it by adding some restrictions such as the inside of C and outside of D must be disjoint to avoid such a counter example [22]. They defined this as strong local digital deformation. However, it is sufficient but will miss some cases. In addition, their method is difficult to deal with in 3D cases.

Herman [20] simply used single surface-cell reduction/addition to help the definition of deformation [20]. This is closer to the original meaning of topology but it also has the same restrictions in 2D images. Herman's concept is called elementarily 1-equivalent. However, Herman's definition is not strictly defined in of mathematics. Boxer [5] modified his own earlier definition [5].

In fact, Newman in 1954 had already discussed similar cases in his seminal book for continuous space [23]. The most important aspect of the theory concerns whether a surface-unit (two-cell) surrounded by a union of two curves is in the "target space" The problem is that in continuous space, we define complexes, even through all point of the boundary are in the "target space," but the face on the "inside" might not be in the target space. This is because if the target space is not a single point, the inside could always contain another point.

In digital cases, since we only have a finite number of points, if we define complexes, it will generate an ambiguity. We have to force the case to be inclusive meaning that if all corner points of a two-cell are in the target space, then the two-cell itself is in the target unit. Chen [10] realized that in order to define a digital curve, one must first define a digital surface [10]. The deformation can be defined on moving a curve along with the different sides of two-cells on the digital surface.

So we can say: Two simple digital curves are gradually varied if the union contains no closed curves. This union can only contain semi-closed curves that are the union of some two-cells [11].

Researchers have also developed some interesting connections about covering space and fundamental groups [19]. It shows that the digital continuous functions potentially have the power to calculate homotopy properties of three-dimensional images. Although the calculation of homotopy groups is an undecidable problem in general, for many real data sets it can be done. This is a very important task for image processing and recognition. For homotopy groups the calculation seems as if it requires "continuous" functions in discrete space. Digitally continuous function can play a major role in terms of digital images. Researchers have proven the linear time algorithm for calculating genus and homology groups. Recently, Chen and Rong have found a simple formula related to the homology groups specifically to calculate the genus of 3D objects [17].

1.6 Smooth Digital Functions and Applications

Digital Functions with smoothness is an interesting topic. First, it is ideal to have smooth looking digital functions, but it is difficult to describe how smooth these digital functions are.

A direct way is to use the finite difference method to define smoothness. The original definition of digitally continuous function could not deal with a smooth function since the largest derivative is "1." In 2010, Chen proposed a two step-method to obtain smooth-looking functions based on sampling data: (1) use the digital method to get the "continuous" data fitting, the method is called gradually varied fitting, (2) use the discrete method to restore the original value to calculate the derivatives of the desired functions [13]. Based on the theory of the new method, a systematic digital-discrete method for obtaining continuous functions with smoothness to a certain order ($C^{(n)}$) from sample data was discovered.

For higher order derivatives, this method alternately uses the digital and discrete methods to obtain the functions needed. We then use the Taylor expansion to combine them. This function can also be called a smooth gradually varied function, referring to its connections to gradually varied functions.

1.6.1 Algorithm Design

The purpose of algorithm design is to find an efficient way to solve a problem. In this book, we want to obtain the smooth digital functions. The algorithm tries to find the best solution to the fitting. We have added a component of the classical finite difference method. The major steps of the new algorithm in 2D include:

(1) Load guiding points. In this step we load the data points with observation values.
(2) Determine resolution. Locate the points in grid space.
(3) Function extension according to the theorem presented in Sect. 1.3. This way, we obtain gradually varied or near gradually varied (continuous) functions.
(4) Use the finite difference method to calculate partial derivatives. Then obtain the smoothed function.
(5) Some multilevel and multi resolution method may be used.

1.6.2 General Digital-Discrete Data Reconstruction and Applications

Real world problems related to data fitting and reconstruction are always at the center of computational sciences and engineering. It affects people's everyday lives ranging from medical imaging, bridge construction, rocket launching, the auto industry, etc. The classical methods are numerical approaches including numerical data fitting and partial differential equations.

This book also deals with a variety of applications using the digital-discrete method. Our overall problem is: Based on finite observations, how do we extend a solution to whole area with continuity and smoothness? It is easy to match its application in the real world. For example, when BP's Macondo well was leaking oil in the Gulf of Mexico, we would like to send a ship to investigate how this area is affected by having the ship pick up random samples. How do we calculate the distribution of the effects? This domain is a rectangular area.

Another example must use a closed surface as its domain. Engineers in the car industry design car shapes, usually with computer aided design (CAD) technology. A designer determines certain values at a few positions on the car body. We would like to have a program that can automatically fit a surface lying on those fix points. The functions on discrete manifolds, especially smooth gradually varied functions, can make considerable contributions to this idea.

This book also make efforts in practical implementations. Selecting good data structures is essential to build a fast algorithm in applications. We provide details for 2D domains first. This part is specific to real data processing in 2D rectangular domains. After that, we discuss to establish a linked-cell data structure for function reconstruction on arbitrary domains and manifolds. We introduce the discrete manifolds (meshes), digital manifolds, and data structures for storing the manifold data. Several algorithms are presented for different applications.

1.6.3 The Digital-Discrete Method and Harmonic Functions

To get a smooth function on a 2D or 3D manifold is a common problem in computer graphics and computational mathematics. In computer graphics, smooth data reconstruction on 2D or 3D manifolds usually refers to subdivision problems. Such a method is only valid based on dense samples. The manifold usually needs to be triangulated into meshes (or patches) and each node on the mesh will have an initial value. While the mesh is refined, the algorithm will provide a smooth function on the redefined manifolds.

When data points are not dense and the original mesh is unchangeable, how do we make a "continuous and/or smooth" reconstruction? In this book, we will present a new method using harmonic functions to solve the problem. Our method contains the following steps:

(1) Partition the boundary surfaces of the 3D manifold based on sample points so that each sample point is on the edge of the partition.
(2) Use gradually varied interpolation on the edges so that each point on the edge will be assigned a value. In addition, all values on the edge are gradually varied.
(3) Use discrete harmonic function to fit the unknown points, i.e. the points inside each partition patch. Finally, we can use the boundary surface to do the harmonic reconstruction for the original 3D manifold.

The fitted function will be a harmonic or local harmonic function in each partitioned area. The function on edge will be "near" continuous (or "near" gradually varied). If we need a smoothed surface on the manifold, we can apply subdivision algorithms.

The new method we present may eliminate some use of triangulations, the foundation of computer graphics for at least 30 years. In the past, people usually used triangulation for data reconstruction. This book employs harmonic functions, a generalization of triangulation, because linear functions are a form of harmonic functions. Therefore, local harmonic initialization is more sophisticated than triangulation.

This new method contains a philosophical change in computer graphics. Triangulation is no longer necessary? It would be a discovery which is incredibly exciting! Moreover, the method we present is the generalization of triangulation and bi-linear interpolation (in Coons surfaces, four points interpolation). It is true that triangulation and bilinear functions are both harmonic. However, harmonic functions do not only refer to linear functions [15, 16].

1.7 Gradually Varied Functions and Other Computational Methods

A general data reconstruction method should be applicable to a variety of real problems including partial differential equations and graphical methods. The gradually varied function-based digital-discrete method has tremendous potential in solving a wide range of partial differential questions in practice.

In this book, we have presented a way to solve parabolic differential equations and also use it in groundwater equations.

For computer graphics and geometric design, the subdivision method and moving least square method are very popular today. We have made some connections from gradually varied functions to these methods. We also deal with some concerns about the relationship between artificial intelligence methods and digital-discrete methods in this book. New applications of digital functions can also be found in [18].

References

1. Agnarsson G, Chen L (2006) On the extension of vertex maps to graph homomorphisms. Discret Math 306(17):2021–2030
2. Alexandrov PS (1998) Combinatorial topology. Dover, New York
3. E. Bishop and D. Bridges (1985) *Constructive Analysis,* Springer Verlag, 1985.
4. Boxer L (1994) Digitally continuous functions. Pattern Recognit Lett 15(8):833–839
5. Boxer L (1999) A classical construction for the digital fundamental group. J Math Imaging Vis 10(1):51–62
6. Chen L (1990) The necessary and sufficient condition and the efficient algorithms for gradually varied fill. Chinese Sci Bull 35:10
7. Chen L (1991) Gradually varied surfaces on digital manifold. In: Abstract of international conference on industrial and applied mathematics, Washington, DC, 1991
8. Chen L (1994) Gradually varied surface and its optimal uniform approximation. In: *IS&T* SPIE symposium on electronic imaging, SPIE proceedings, vol 2182. San Jose. (L. Chen, Gradually varied surfaces and gradually varied functions, in Chinese, 1990; in English 2005 CITR-TR 156, University of Auckland. Has cited by IEEE Trans in PAMI and other publications)

9. Chen L, Cheng HD, Zhang J (1994) Fuzzy subfiber and its to seismic lithology classification. Inf Sci 1(2):77–95
10. Chen L (1999) Note on the discrete jordan curve theorem. Proceedings of the SPIE on Vision geometry VIII, vol 3811. SPIE, Denver
11. Chen L (2004) Discrete surfaces and manifolds: a theory of digital-discrete geometry and topology. Scientific and Practical Computing, Rockville, MD
12. Chen L, Gradual variation analysis for groundwater flow of DC (revised). DC Water Resources Research Institute Final Report, Dec 2009. http://arxiv.org/ftp/arxiv/papers/1001/1001.3190. pdf
13. Chen L (2010) A digital-discrete method for smooth-continuous data reconstruction. J Wash Acad Sci 96(2):47–65 (ISSN 0043-0439). http://arxiv.org/ftp/arxiv/papers/1010/1010.3299. pdf
14. Chen L (2010) Digital-discrete surface reconstruction: a true universal and nonlinear method. http://arxiv.org/ftp/arxiv/papers/1003/1003.2242.pdf
15. Chen L, Liu Y, Luo F (2009) A note on gradually varied functions and harmonic functions. http://arxiv.org/PS_cache/arxiv/pdf/0910/0910.5040v1.pdf
16. Chen L, Luo F (2011) Harmonic functions for data reconstruction on 3D manifolds, Submitted for publication. http://arxiv.org/ftp/arxiv/papers/1102/1102.0200.pdf
17. Chen L, Rong Y (2010) Digital topological method for computing genus and the betti numbers. Topol Appl 157(12):1931–1936
18. Escribano C, Giraldo A, Sastre M.A (2012) Digitally continuous multivalued functions, morphological operations and thinning algorithms, J Math Imaging Vis 42(1):76–91
19. Han SE (2005) Digital coverings and their applications. J Appl Math Comput 18(1–2):487–495
20. Herman GT (1993) Oriented surfaces in digital spaces. CVGIP: Gr Model Image Process 55:381–396
21. Khalimsky E (1987) Motion, deformation, and homotopy in finite spaces. In: Proceedings IEEE international conference on systems, man, and cybernetics, pp 227–234, Chicago
22. Kong TY (1989) A digital fundamental group. Comput Graph 13:159–166
23. Newman M (1954) Elements of the topology of plane sets of points. Cambridge, London
24. Rosenfeld A (1986) Continuous' functions on digital pictures. Pattern Recognit Lett 4:177–184
25. Rosenfeld A (1996), Contraction of digital curves, University of Maryland's Technical Report in Progress. ftp://ftp.cfar.umd.edu/TRs/trs-in-progress/new.../digital-curves.ps
26. Rosenfeld A, Nakamura A (1997) Local deformations of digital curves. Pattern Recognit Lett 18:613–620

Chapter 2
Functions and Relations

Abstract This chapter overviews the basic concepts of functions and relations including continuous functions and their differentiations in Euclidean space. We also introduce functions in discrete spaces, specifically graphs and grid spaces. This chapter contains three parts: continuous functions in Euclidean space, graphs and discrete spaces, and advanced topics including topological spaces. Philosophically, a space is a relation or a collection of relations over a set. A graph is a relation; an m-dimensional Euclidean Space is a collection of relations over points in R^m. This chapter is written in a gradual progression to provide a simple and interesting review of the background knowledge. Some of the more profound issues will be presented in later chapters as needed. Those with a strong background in mathematics can skip some or all of this chapter. This chapter prepares the necessary basic knowledge for the rest of the book.

2.1 Definitions of Functions and Relations

Functions and relations are the most popular mathematical terms in science. For instance, $f(x) = x^2 + 1$ is a function on real number set R. A function is a mapping from a set to another. One usually writes $f : R \to R$, and the polynomial $x^2 + 1$ indicates the way to obtain the values of the function.

A relation is a collection of pairs. For example $\mathbf{R} = \{(x, x^2)|x \in R\}$ is a relation on R.

In general, for two sets A and B, a function maps each element of A to an element of B. A relation can correspond one element in the first set to multiple elements in the second set. For functions, the first set is usually called the domain, the second set is called the range or codomain. The range of functions usually means real numbers. For relations, it is not necessary for every element of the domain to have a corresponding member in the codomain.

The range of a function sometimes also means the image of the function, which indicates the subset of actual values taken from the codomain.

L.M. Chen, *Digital Functions and Data Reconstruction: Digital-Discrete Methods*, 13
DOI 10.1007/978-1-4614-5638-4_2, © Springer Science+Business Media, LLC 2013

2.1.1 Functions on Sets

In mathematics, a set is a collection of elements and can be any object. For example, $S = \{a,b,c\}$ can be a set of three balls a, b and c. We also use $x \in S$ to indicate that x is any member (also called element) of S. Let S be a set. S' is a subset of S meaning that each element of S' is an element of S. Their relationship is denoted by $S' \subseteq S$. If S is not a subset of S', then S' is called a proper-subset of S, denoted by $S' \subset S$ [8]. Three basic operations about sets are listed below:

(a) Intersection: $A \cap B = \{x | x \in A \text{ and } x \in B\}$;
(b) Union: $A \cup B = \{x | x \in A \text{ or } x \in B\}$;
(c) Complement: $A - B = \{x | x \in A \text{ and } x \notin B\}$;

The formal definition of a function can be described as a mapping f from Set X to Set Y, denoted $f : X \to Y$. A function maps each element of X to exactly one element of Y.

A function is called a *one-to-one* function if no two elements in X map to the same element in Y. A function is called *on-to* if all elements in Y have been mapped, meaning that for $y \in Y$, there exists an $x \in X$, such that $f(x) = y$. A function f that is both one-to-one and onto is called invertible, its inversion is denoted by f^{-1}.

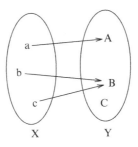

Fig. 2.1 A function from set A to B

2.1.2 Concept of Relations

In order to provide a relatively precise description of relations, we need to define the ordered pair. Let A and B be two sets where $a \in A$ and $b \in B$. (a,b) is said to be an ordered pair from A to B. Two ordered pairs (a,b) and (a',b') are equal if and only if $a = a'$ and $b = b'$.

R is said to be a binary relation from a set A to a set B if it is formed by a collection of ordered pairs (a,b) where $a \in A$ and $b \in B$.

In set theory, the Cartesian product of A and B, $A \times B$, is the set of all ordered pairs (a,b). Therefore, a binary relation is a subset of Cartesian product $A \times B$. We also define the Cartesian power of a A as:

$$A^n = A \times A \times \cdots A = \{(a_1, a_2, \cdots, a_n) | a_i \in A_i \text{ for all } 1 \le i \le n\} \qquad (2.1)$$

An important concept called the equivalence relation is defined as when $A = B$ (or a relation over A), $\mathbf{R}$ satisfies the following three conditions:

(a) Reflexive: for all $a \in A$, $(a,a) \in \mathbf{R}$.
(b) Symmetric: if $(a,b) \in R$ then $(b,a) \in \mathbf{R}$.
(c) Transitive: if (a,b) and $(b,c) \in \mathbf{R}$ then $(a,c) \in \mathbf{R}$.

In addition, a relation that is reflexive and symmetric is called a similarity relation. A relation that is reflexive, asymmetric, and transitive is called a partial order (a tree is a partial order). A partial order is said to be a total order if for a and b, $a \neq b$, $(a,b) \in \mathbf{R}$ or $(b,a) \in \mathbf{R}$. Therefore, a chain is a total order.

A graph $G = (V,E)$ can be defined using a relation. A directed graph can be defined as an ordered relation and an undirected graph can be defined as a symmetric relation E over vertices V. See Sect. 2.3.

2.2 Continuous Functions in Euclidean Space

Continuous functions are usually defined in Euclidean space. The intuitive meaning of the continuous function f is that there are no jumps or gaps in the function. In mathematics, it is defined using *limits*. For short, let δ be a small positive number or interval. f is continuous at a point x_0 if

$$f(x_0) \approx f(x_0 + \delta) \approx f(x_0 - \delta). \tag{2.2}$$

In other words, $f(x)$ is said to be continuous at x_0 if

$$f(x_0) = \lim_{\delta \to 0} f(x_0 + \delta) = \lim_{\delta \to 0} f(x_0 - \delta). \tag{2.3}$$

If $f(x)$ is continuous at every point in the domain, then $f(x)$ is continuous. A typical example of a continuous function is shown in Fig. 2.2.

In calculus and functional analysis, the *Lipschitz* condition is a special form of continuity for functions. If there exists a constant L such that

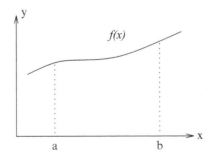

Fig. 2.2 A continuous function on R

$$|f(x) - f(y)| \leq L \cdot |x - y| \tag{2.4}$$

for all x and y, then f is called a Lipschitz function while L is called a Lipschitz constant. For instance, any bounded function is the Lipschitz function. A function f is said to be local Lipschitz if for every x there exists a neighborhood U_x of x such that f on U_x is a Lipschitz function. In other words, f is local Lipschitz if for any x, y and $|x - y| \leq d$ (another description of U_x), there exists $L_{x,d}$ such that

$$|f(x) - f(y)| \leq L_{x,d} \cdot |x - y| \tag{2.5}$$

Most of the continuous functions are local Lipschitz functions. In this book, we assume all functions are local Lipschitz.

2.2.1 Differentiation and Integration of Functions

Differentiation and integration are two basic concepts in calculus [9]. The differentiation of a function is introduced by the derivatives as defined below:

$$f'(x) = \lim_{y \to x} \frac{f(y) - f(x)}{y - x} \tag{2.6}$$

If $f'(x)$ is continuous, then $f(x)$ is said to be differentiable or differentiable in the first order. We can further define nth order differential functions. A function is said to be infinitely differentiable if it has any order of differentiation. We also call a function to be smooth if it is infinitely differentiable.

We use C^0 to represent the class that consists of all continuous functions. The class C^k consists of all differentiable functions whose kth derivative is continuous. C^∞ represents an infinitely differentiable class.

We usually use $\frac{dy}{dx}$ or $\frac{\Delta y}{\Delta x}$ to represent $f'(x)$ where dy and dx are symbolic representations of differentiation. They do not have an exact numerical meaning and only represent small valued elements.

The Taylor series expansion of a function f is a polynomial around point x_0:

$$T_f(x) = f(x_0) + \frac{1}{1!} f^{(1)}(x_0)x + \cdots + \frac{1}{n!} f^{(n)}(x_0)x^n + \cdots \tag{2.7}$$

A function f is said to be analytic if f is in C^∞ and equals its Taylor series expansion around any point. We mainly deal with analytic functions in this book. In other words, we assume $f = T_f(x)$ around any point x_0.

The integral of a function, meaning the inverse function of a derivative called the antiderivative, is denoted as

$$F = \int f(x) \, dx. \tag{2.8}$$

In other words, the derivative of F is f: $F' = f$.

A major achievement in mathematics is called the principles of integration, the Newton-Leibniz formula , or the Fundamental Theorem of Calculus . This formula states that for a piece-wise continuous function (meaning a function that only contains finite discontinuities), the area bounded by f and x-axis line from point a to b (See Fig. 2.2), denoted as a definite integral $\int_a^b f(x)\,dx$, is $F(b) - F(a)$, i.e.

$$F(b) - F(a) = \int_a^b f(x)\,dx \tag{2.9}$$

For a continuous function, the Weierstrass approximation theorem is fundamental to science and engineering. It is basic to this book: Every continuous function on an interval $[a, b]$ can be uniformly approximated by a polynomial function.

The Bernstein polynomial is such a polynomial function. It is also used in computer graphics, called Bezier Polynomial. We will present more on this topic in Chap. 6.

2.2.2 Finite Differences and Approximations of Derivatives

Finite difference Method is a main technology for the numerical solution of many problems involving derivative and differential calculations. Finite differences are very straightforward and use small changes to replace limits.

Since Δx can be viewed as a small constant, we let $\Delta x = h$. Then

$$\Delta f = f(x+h) - f(x) \tag{2.10}$$

is called the first order difference, and we have

$$f'(x) \approx \frac{\Delta f}{\Delta x} = \frac{f(x+h) - f(x)}{h} \tag{2.11}$$

Based on

$$f''(x) \approx \frac{\Delta^2 f}{\Delta x^2} \tag{2.12}$$

we can retrieve the second order difference:

$$\Delta^2 f = \Delta(\Delta f) = (f(x+h) - f(x)) - (f(x) - f(x-h)) = f(x+h) - 2f(x) + f(x-h) \tag{2.13}$$

For many applications, h might be fixed as "1." For a better approximation, the other formats may be chosen. For instance,

$$f'(x) = -\frac{f(x+2h) - 4f(x+h) + 3f(x)}{2h}$$

more related formulas will be presented in Chaps. 7, 9, and 12.

2.3 Graphs and Simple Graphs

In this section, some basic concepts in graph theory are reviewed. A graph G consists of two sets V and E, where V is a set of vertices and E is a set of edges formed by pairs of vertices. An edge is said to be incident with a vertex it joins [6].

$G = (V,E)$ is called an undirected graph if $(a,b) \in E$ then $(b,a) \in E$, (or simply $(a,b) = (b,a)$). G is said to be a simple graph if every pair of vertices has only one edge incident to the two vertices and there is no $(a,a) \in E$ for any $a \in V$. In this book, we assume that $G = (V,E)$ is a undirected and simple graph, unless otherwise noted. See Fig. 2.3(a).

If (p,q) is in E, p is said to be adjacent to q. Let $p_0, p_1, \ldots, p_{n-1}, p_n$ be $n+1$ vertices in V. If (p_{i-1}, p_i) is in E for all $i = 1, \ldots, n$, then $\{p_0, p_1, \ldots, p_{n-1}, p_n\}$ is called a path . If $p_0, p_1, \ldots, p_{n-1}, p_n$ are distinct vertices, the path is called a simple path . The length of this path is n .

A simple path $\{p_0, p_1, \ldots, p_{n-1}, p_n\}$ is closed if (p_0, p_n) is an edge in E [4, 15]. A closed path is also called a cycle. Two vertices p and q are connected if there exists a path $\{p_0, p_1, \ldots, p_{n-1}, p_n\}$ such that $p_0 = p$ and $p_n = q$. G is called connected if every pair of vertices in G is connected. In this book, it is always assumed that G is connected.

The distance $d(x,y)$ between two vertices x and y is the length of the shortest path between them. For a measure $d(x,y)$, we have $d(x,y) \leq d(x,a) + d(a,y)$. This is called the *triangle inequality* [8]. This is also true for real numbers or vectors: $|x+y| \leq |x| + |y|$, and $|x-y| \leq |x| + |y|$.

Suppose $G' = (V',E')$ is a graph where $V' \subset V$ and $E' \subset E$ for graph $G = (V,E)$, $G' = (V',E')$ is called a subgraph of G. If E' consists of all edges in G whose joining vertices are in V', then the subgraph $G' = (V',E')$ is called a partial-graph of G and their relationship is denoted by $G' \preceq G$. If V' is a proper-subset of V, then the relationship is denoted by $G' \prec G$.

It is noted that for a certain subset V' of V, the partial-graph G' with vertices V' is uniquely defined.

The planar graph is a graph can be drawn in a plane without crossing any edges.

2.4 Space, Discrete Space, Digital Space

A space can be viewed as a collection of relations on a base set. Therefore, graphs are spaces. An n-dimensional Euclidean space is the n-dimensional real space R^n, the Cartesian power of R, with the distance metric relation $d(x,y) = \sqrt{((x_1 - y_1)^2 + \cdots + (x_n - y_n)^2)}$. This is the popular Euclidean metric.

Euclidean space is the continuous space. Other examples of continuous spaces include the circle and sphere. These spaces refer to manifolds. We will introduce this concept in the next section.

The simplest discrete space is called the grid space that contains all integer points in Euclidean space. For instance, the three dimensional (3D) grid space is

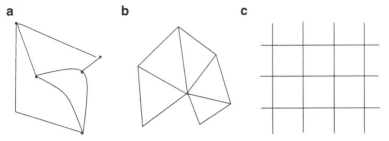

Fig. 2.3 Examples of Graphs: (**a**) Graphs, (**b**) Triangulated graphs, (**c**) 2D Grid graphs

$$G = \{(x,y,z)|x,y,z \in I, I \text{ is the set of integers}\}$$

If we select a metric or distance measure, then we will have a digital space. The major difference between continuous space and digital space is that in digital space, the distance metric is usually selected as a discrete metric and not the Euclidean distance.

The general discrete space is described by a graph associated with a metric. The most widely used discrete space is triangulation, formally called simplicial complexes. See Fig. 2.3b. Digital space refers to grid space. See Fig. 2.3c.

2.4.1 Grid Space and Digital Space

In this book, Σ_m represents a special graph $\Sigma_m = (V,E)$. V contains all integer grid points in m dimensional Euclidean space. The edge set E of Σ_m is defined as $E = \{(a,b)|a,b \in V \text{ and } d(a,b) = 1\}$, where $d(a,b)$ is the Euclidean distance between a and b [3].

In fact, E contains all pairs of adjacent points. Let $a = (x_1,\ldots,x_m)$ and $b = (y_1,\ldots,y_m)$. Because a is an m-dimensional vector, $(a,b) \in E$ means that there is only one component i such that $|x_i - y_i| = 1$ and the rest of the components in a and b are the same. This is known as direct adjacency, i.e. $\sum_{i=1}^{n} |x_i - y_i| = 1$. One can define indirect adjacency as $\max_{i=1}^{m} |x_i - y_i| = 1$. See Fig. 2.4.

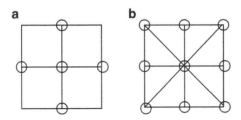

Fig. 2.4 Direct and indirect adjacencies in 2D: (**a**) four-adjacency, and (**b**) eight-adjacency

Fig. 2.5 Cells in digital space: (**a**) one-cell, (**b**) two-cell, (**c**) three-cell

Σ_m is usually called an m-dimensional digital space . The basic discrete geometric element n-dimensional cells (nD-cells, or n-cells) can be defined in such a space, including zero-cells (point-cells), one-cells (line-cells), and two-cells (surface-cells) in Fig. 2.5. We will discuss more of this in Chap. 8. A three-cell is also called a cube, so Σ_3 with direct adjacency (six-adjacency) is called the cubic space.

2.4.2 Triangulation

Discrete geometry usually deals with a finite number of objects and their relationships in Euclidean space. Triangulation means to partition a domain into triangles. It is the major methodology in discrete geometry. Triangulation can be used to split a domain into triangles and to represent part of a 3D object with a small 2D element [5, 7].

The advantage of using triangles as opposed to rectangles is the flexibility of the combination of different triangles. It would be sometimes very difficult to represent irregular domains using squares or grids. See Chap. 6.

Given n points in a 2D plane, trianglizing the plane can be done easily (See Fig. 2.3b). However, there are too many of these types of triangulations. If one use the triangulation for data reconstruction, the different triangulations would result in different piecewise interpolations.

2.4.3 Polygons and Lattice

A more general type of triangle and rectangle is called the polygon. A polygon refers to a flat shape consisting of line segments and vertices. A polygon can be viewed as a simple closed path in graph theory. However, polygons have geometric distance metric. A convex polygon means for two points in the polygon, the linking line between them is also completely inside of the polygon. Triangles and rectangles are convex polygons [5, 7].

A 2D domain can be decomposed into a set of polygons. We sometime say it forms lattice graphs or meshes. This name is used ubiquitously in computer graphics. Note that in mathematics, a lattice may refer to a binary relation. An algebraic lattice is a partial order relation in that for any two elements there exists a unique least upper bound and a unique greatest lower bound.

The Voronoi diagram decomposes a 2D domain into polygons based on n sample points. In Voronoi diagrams, each polygon contains a sample point, also called a site. A point x is inside a specific polygon containing site p if x is closer to p than to any other site.

Each edge in a Voronoi polygon, not intersecting with the boundary of the domain, will be shared by two polygons. Edges are the linking line between the two sites of these two polygons; when we generate triangulations of the domain, it is called Delaunay triangulation. Therefore, Delaunay triangulation is a dual diagram of the Voronoi diagram.

Delaunay triangulation is the most popular form among different types of triangulation. The Voronoi diagram has many applications in the real world too. We will give algorithms to obtain them in Chap. 6. Figure 2.6 shows a real data example computed by D. Mount.

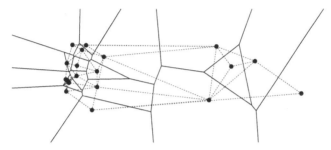

Fig. 2.6 Delaunay triangulation and Voronoi diagram

2.5 Functions on Topological Spaces*

To understand a function on a manifold or general topological space is much harder than understanding it as defined on a plane or 3D Euclidean space. A good example is to think about the smoothed color painting on the surface of a car. We usually need to deal with a smooth function (colors) on a smooth manifold (car shapes).

Although we do not need the perfect mathematical definition of topological space in this book, in order to maintain a degree of self-containment, we would like to include some brief definitions here. The definition of a topological space is very general. It is based on the definition of topology on sets [1, 2].

Definition 2.1. Let X be a set and τ be a collection of subsets of X. (X, τ) is said to be a topological space if (X, τ) satisfies the following axioms: (1) The empty set and X are in τ. (2) The union of arbitrary number of elements of τ is in τ. (τ is closed under arbitrary union.) (3) The intersection of any finite number of elements of τ is in τ. (τ is closed under finite (number of) intersection.)

In mathematics, τ is called a topology on X. The elements of X are called points, the elements of τ are called open sets. The complement of an open set A in X, $X - A$, is called a closed set.

Almost all geometric objects or spaces we deal with are topological spaces. For a finite set X, the open set is the same as the close set. A topological space can also refer to a function space in which each element of X is a function.

Functions on topological spaces usually mean that the function is on the base set X. We can also define a function between two topological spaces (X, τ) and (Y, τ'). For instance, $f : X \to Y$.

Intuitively, we say that two objects are topologically equivalent if there is a process that can continuously change an object into another. It can be defined as a continuous one-to-one onto function. We also say that these two objects have homeomorphism.

Definition 2.2. (X, τ) and (Y, τ') are said to be homeomorphic or topologically equivalent if there exists a continuous and invertible function f.

A (topological) n-manifold is a topological space $M = (X, \tau)$. Each of element (point) of X has an open nD neighborhood U_x that is continuously equivalent or homeomorphic to an nD Euclidean space, E_n. Homeomorphism means that there is a continuous function $f_x : U_x \to E_n$ where f_x^{-1} is also continuous.

A smooth n-manifold is a manifold where for any two open sets U_x and U_y in M, $f_y \cdot f_x^{-1}$ on $f_x(U_x \cap U_y)$ is smooth or C^k-continuous. Smoothness is defined on E_n, but through M. The intuitive meaning of this definition is that the local homeomorphic functions f_y and f_x on intersection $U_x \cap U_y$ guarantees that it can be smoothly put on $U_x \cap U_y$ without creating a bent angle in each local space.

References

1. Alexandrov PS (1998) Combinatorial topology. Dover, New York
2. Armstrong MA (1997) Basic topology, Rev edn. Springer, New York
3. Chen L (2004) Discrete surfaces and manifolds: a theory of digital-discrete geometry and topology. Scientific and Practical Computing, Rockville, MD
4. Cormen TH, Leiserson CE, Rivest RL (1993) Introduction to algorithms. MIT, Cambridge
5. Goodman JE, O'Rourke J (1997) Handbook of discrete and computational geometry. CRC, Boca Raton
6. Harary F (1972) Graph theory. Addison-Wesley, Reading
7. Preparata FP, Shamos MI (1985) Computational geometry: an introduction. Springer, New York
8. Rosen KH (2007) Discrete mathematics and its applications. McGraw-Hill, New York
9. Thomas GB Jr (1969) Calculus and analytic geometry, 4th edn. Addison-Wesley, Reading

Chapter 3
Functions in Digital and Discrete Space

Abstract In this chapter, we introduce the concepts of digital functions and their basic properties. We start with digital "continuous" functions by Rosenfeld, then on to gradually varied functions proposed by Chen. We also present the necessary and sufficient condition of existence for gradually varied interpolations. For practical uses of these concepts, algorithm design is also a central topic in this chapter. We have designed different algorithms for gradually varied extension. Digital continuous functions and gradually varied functions are similar, but have some differences. Digital continuous functions are mainly defined on digital spaces and gradually varied functions are defined on any discrete space. At the end of the chapter, we discuss the mathematical foundations for gradually varied functions through constructive analysis.

3.1 What is a Digital Function

A digital function is a function mainly defined on digital space. Digital space can be viewed as a subspace of Euclidean space. This subspace contains all integer grid points. We usually use Σ_m to represent an m-dimensional digital space. Digital space is usually limited in each dimension in terms of length. Sometimes $I^{(m)}$ is regarded as the m-dimensional integer grid space with unlimited length in each dimension.

As it was presented in Chap. 2, a connectivity (or relation) must be defined over Σ_m before it can be treated as a digital space. For example, four-connectivity and eight-connectivity are two options for 2D. See Fig. 2.4. This is because a digital point p has four direct adjacent digital points as its neighbors in 2D. Also, p can have eight neighbors if we count diagonal neighbors as well [16]. For the same reason, 6, 18, and 26-connectivity are for 3D digital spaces.

A digital function is a function from a digital space to I, the integers. We usually take the set $\{1, 2, \cdots, n\}$ as the function's range. For instance, $f : \Sigma_2 \to \{1, 2, 3, \cdots, n\}$ is a digital function. Beyond this narrow definition, a broader meaning for the digital function can be viewed as a function in general discrete space. It can also be a mapping for two computerized objects.

L.M. Chen, *Digital Functions and Data Reconstruction: Digital-Discrete Methods*,
DOI 10.1007/978-1-4614-5638-4_3, © Springer Science+Business Media, LLC 2013

The study of digital spaces has a long history, which can be traced back to geometry of numbers by Hermann Minkowski more than 100 years ago. Minkwoski discovered an important theorem stating that an integer grid point must be involved in a convex region for it to have sufficient volume [13].

The modern development of this idea is largely with respect to image processing and computer vision. A digital image or picture is a typical example of the function on 2D digital space. In order to find an object in an image, we usually need to partition the image into several connected components. This process is called image segmentation. Rosenfield defined digital continuous functions for such a purpose.

3.2 Digital Continuous Functions

A digital continuous function is defined as a function in which the integer values of its digital points is the same or off by at most "1" from the values of its neighbors. Mathematically,

Definition 3.1. Let $f : \Sigma_k \to \{\cdots, 1, 2, 3, \cdots\}$ be a digital function. f is said to be a digital continuous function if for any adjacent points x and y, $|f(x) - f(y)| \leq 1$.

In Fig. 3.1a, the function value jumps from 2 to 4 when $x = 2$ moves to $x = 3$. Therefore, this function is not "continuous." The function in Fig. 3.1b is a digital continuous function.

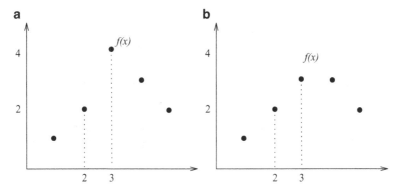

Fig. 3.1 Examples of Digital Functions: (**a**) a digital function but not "continuous"; (**b**) a digital continuous function

Rosenfeld (1986) was the first to mathematically define a digital continuous function on Σ_k [15]. Properties of digital continuous functions are also valid when the Lipschitz condition is naturally satisfied. Since the domain Σ_k is discrete space, we use $d(x, y) = |x - y|$ to represent the length from x to y. Suppose $P(x, y)$ is a path with the shortest distance between x and y, $P(x, y) = \{x_0, x_1, \cdots, x_n\}$, $x = x_0$, and $x_n = y$.

Then, $d(x,y) = n$. By the definitions of digital continuous function and triangle inequality, we have $f(x_0) - f(x_n) = f(x_0) - f(x_1) + f(x_1) - f(x_2) + \cdots + f(x_{n-1}) - f(x_n) = \sum_{i=0}^{n-1}(f(x_i) - f(x_{i+1}))$. Thus $|f(x_0) - f(x_n)| \leq \sum_{i=0}^{n-1}|(f(x_i) - f(x_{i+1}))| \leq \sum_{i=0}^{n-1}(1) \leq n$. Therefore [15],

Proposition 3.1. *Any digital continuous function is a Lipschitz function, i.e.* $|f(x) - f(y)| \leq d(x,y)$.

Because every pair of adjacent points differs by at most 1, we can immediately obtain the following:

Proposition 3.2. *If f is a digital continuous function on Σ_k, the values f on any path of Σ_k is also digital continuous. Therefore, f is digital continuous on any connected component of Σ_k.*

If we use the finite difference method for calculating the derivative of a digital continuous function, there are only three possible values of derivatives, 0, -1, 1. Conversely, a digital continuous path with respect to f might not guarantee that we can get a digital continuous region based on the path. We will present a related theorem in the next section.

3.3 Gradually Varied Functions

The gradually varied function was introduced and studied in digital and discrete spaces by Chen in 1989 [3]. The basic idea of introducing gradually varied functions is to use a purely discrete interpolation algorithm to fit a discrete surface when the desired surface is not required to be "smooth."

Let $A_1, A_2, \ldots, A_m$ be the m rational or real numbers where $A_1 < A_2 < \ldots < A_m$. Assume f is a function from Σ_2 (or any other discrete space) to $\{A_1, A_2, \ldots, A_m\}$. For two points p and q in Σ_2, if $f(p) = A_i$ and $f(q) = A_j$, then the *level-difference* between $f(p)$ and $f(q)$ is $|i - j|$. We define the gradually varied function below:

Definition 3.2. Let p, q be two adjacent points in Σ_2. f is said to be gradually varied on p and q, if $f(p) = A_i$ implies $f(q) = A_{i-1}, A_i$, or A_{i+1}.

Definition 3.3. f is said to be gradually varied if f is gradually varied on any pair of adjacent points p, q in Σ_2.

In general, Let $G = (V, E)$ be a graph. A function $f : G \to \{A_1, A_2, \cdots, A_m\}$ is gradually varied if for any adjacent pair p, q in G, $f(p) = A_i$ implies $f(q) = A_{i-1}$, A_i, or A_{i+1}.

We can see that gradually varied is more general than digitally continuous, introduced in the above section, since we can let $A_i = i$ for all $i = 1, 2, \cdots, m$. The problem with gradually varied interpolation is as follows:

Let D be a connected subset in Σ and $J \subset D$. If given $f_J : J \to \{A_1, A_2, \ldots, A_m\}$, is there an extension of f_J, $f_D : D \to \{A_1, A_2, \ldots, A_m\}$ such that for all $p \in J$, $f_J(p) = f_D(p)$? The main theorem of this book was proven by Chen in 1989 [3].

Theorem 3.1 (Chen, 1989, 1991). *The necessary and sufficient condition under which there exists the gradually varied interpolation is that for any two points p and q in J, the length of the shortest path between p and q in D is not less than the level-difference between $f(p)$ and $f(q)$.*

Proof. We give a constructive proof here. The basic idea of the construction is to assign a value to a point q that has not been assigned a value, but has a neighbor p, which is a sample point or has already been assigned a value.

In order to make the proof clear, we define $LD(p, p')$ as the level difference between $f(p)$ and $f(p')$: Let $f(p) = A_i$ and $f(p') = A_j$, $LD(p, p') = |j - j|$.

As usual, $d(p, p')$ denotes the length of the shortest path between p and p' in D.

1. First, we will prove the necessary condition. Suppose f is the gradually varied function on D, then f is gradually varied on every path in D. The path may be the shortest path between two points p and p' in J. Hence, the length of the shortest path is not less than the difference of the gray level of p and p', i.e. $d(p, p') \geq LD(p, p')$. (See Proposition 3.2).
2. Second, we will prove the sufficient condition. Suppose we have $f_J : J \rightarrow \{A_1, A_2, \ldots, A_m\}$ and for all $p, p' \in J$, $d(p, p') \geq LD(p, p')$ (in D).

   First let $f_D(p) = f_J(p)$ if $p \in J$ and $f_D(p') = \theta$ if $p' \in D - J$. Define

   $$D_0 = \{p | f_D(p) \neq \theta, p \in D\}$$

Now, $D_0 = J$.

This constructive proof is the following process: If $D_0 \neq D$, we can find a vertex (or point) $r \in D_0$ so that r has an adjacent point x not in D_0, i.e. $x \in D - D_0$. We can assume $f_D(r) = A_i$.

Then, let $f_D(x) = f_D(r) = A_i$, and denote

$$m(x) = \{p | f_D(p) < A_i, p \in D_0\};$$

$$M(x) = \{p | f_D(p) > A_i, p \in D_0\}.$$

There will be three cases:

(i) If there is a $p \in m(x)$, such that $d(x, p) < LD(x, p)$, we know $d(r, p) \leq d(r, x) + d(x, p)$ and $d(r, x) = 1$ (r and x are adjacent points). Therefore, $d(r, p) \leq 1 + d(x, p)$; we have $d(x, p) \geq d(x, p) - 1$. We also know $d(r, p) \geq LD(r, p)$ since $r, p \in D_0$. Thus, $d(x, p) \geq d(r, p) - 1 \geq LD(r, p) - 1$. In addition, $LD(r, p) = LD(x, p)$ since $f_D(x) = f_D(r) = A_i$. Therefore, $d(x, p) \geq LD(r, p) - 1 \geq LD(x, p) - 1$.

According to the assumption $d(x, p) < LD(x, p)$, so $d(x, p) = LD(x, p) - 1$. For any $q \in M(x)$, we have $f_D(p) < f_D(x) < f_D(q)$, so $LD(p, q) = LD(p, x) + LD(x, q)$. Again, $d(p, x) + d(x, q) \geq d(p, q) \geq LD(p, q)$, then

$$d(p, x) + d(x, q) \geq LD(p, x) + LD(x, q) \geq LD(p, x) + 1 + LD(x, q)$$

Hence,

$$d(x,q) \geq 1 + LD(x,q).$$

Therefore, we have proven that if there is a $p \in m(x)$ such that $d(x,p) < LD(x,p)$, then

$$\forall p \in m(x)(d(x,p) \geq LD(x,p) - 1) \text{ and } \forall q \in M(x)(d(x,p) \geq LD(x,q) + 1).$$

Thus, we can modify the valuation of x, from $f_D(x) = A_i$ to $f_D(x) = A_{i-1}$. Obviously, for the new value at x, we have,

$$d(y,x) \geq LD(y,x), \text{ if } y \in D_0 \& f_D(y) \neq A_i,$$

and

$$d(y,x) \geq LD(y,x) = 1, \text{ if } y \in D_0 \& f_D(y) = A_i.$$

(ii) If there exists a $q \in M(x)$ such that $d(x,q) < LD(x,q)$, similar to (i), we reassign $f_D(x) = A_{i+1}$. We obtain $\forall y \in D_0(d(x,y) \geq LD(x,y))$.

(iii) If (i) and (ii) are not satisfied, then $f_D(x) = A_i$ was the correct assignment. Let $D_0 \leftarrow D_0 \cup \{x\}$, then for every pair p and p' in D_0, we have

$$d(p,p') \geq LD(p,p').$$

We repeat above process until $D_0 = D$. After all the points in $D - J$ are valued, we can see that f_D is gradually varied. This is because if x,y are adjacent, then $d(x,y) = 1 \geq LD(x,y)$. So $LD(x,y)$ must be 0 or 1. That is, f_D is gradually varied on every pair of adjacent points. Therefore, f_D is gradually varied on D. □

In the proof above, assuming $p \in J$ and its neighbor $q \notin J$, the construction procedure is to let $f(q) = f(p)$ first and then check whether or not $J \cup \{q\}$ (with the new value) satisfies the gradually varied condition. If it does, then keep the $f(q)$; otherwise, subtract or add a level to $f(q)$. Repeating the above process will fill in all unknown value points in D. In the case study section of this chapter, we first give an example to simulate this procedure.

3.4 The Algorithms for Gradually Varied Functions

The purpose of algorithm design is to get the actual procedure quickly. It is at the center of computer science [1, 12]. The algorithm is defined as a sequence of steps needed to complete a task. The time complexity of an algorithm is the total operations needed during the running time of an algorithm. The time complexity is based on the size of the input of an algorithm. For instance, if we have a 5×5 array, the input size is 25, so the time complexity is a function of input size. We say a binary search algorithm is in $O(\log n)$ time, meaning that the input is n numbers in an ascending or descending order, and the time spent running the algorithm is $c \cdot \log_2(n)$, where c is constant.

In this section, we will introduce three algorithms. We first present a basic algorithm for interpolating a gradually varied surface. This algorithm will be used in a case study that will give an example in each step. Next, we will consider using faster algorithm technology to get an $O(n \log n)$ algorithm for the domain with the Jordan property (a closed curve separates a domain into two components) [10, 11]. For more general discrete geometry structures, refer to [8, 11]. At the end of the section, we will give the algorithm Euclidean space.

3.4.1 A Simple Algorithm

In the proof of Theorem 3.1, we have already presented a basic procedure for gradually varied function reconstruction based on the sample point set J.

We can first assume that we have $d(x,y)$ for all $x,y \in D$. $d(x,y)$ can be obtained using algorithms in graphs [12].

Algorithm 3.1 (Simple Algorithm). Let f_J be a function defined on a non-empty subset J of D. If for some $n > 0$, then:

Step 1: Test whether for all p and p' in J, $d(p,p') \geq LD(p,p)$. If not, then there is no gradually varied function interpolation. (We may fit a best uniform approximation) $D_0 \leftarrow J$.

Step 2: Choose x from $D - D_0$ such that x has an adjacent vertex r in D_0. Assume $f_D(r) = A_i$.

Step 3: Let $f_D(x) = f_D(r) = A_i$. Check x against every vertex p in D_0: If there is a $p \in D_0$ where $(d(x,p) < LD(x,p)$, we will change $f_D(x)$ to A_{i-1} when $f_D(p) < A_i$ or change $f_D(x)$ to A_{i+1} when $f_D(p) > A_i$.

Step 4: Let $D_0 \leftarrow D_0 \cup \{x\}$.

Step 5: Repeat steps 2–4 until $D_0 = D$.

For a graph G, the Floyd-Warshall algorithm can be used to calculate the distances between every pair of vertices in D. The time complexity of this algorithm is $O(n^3)$ [12]. We can assume that we know $d(x,y)$ for all x,y.

Theorem 3.2. *The simplest algorithm (Algorithm 3.1) for gradually varied functions has time complexity $O(n^2)$, if $d(x,y)$ for all x,y in D are known.*

Proof. First the correctness of the algorithm was guaranteed by the proof of Theorem 3.1. The algorithm for obtaining the distance between each pair in a graph $G = (V,E)$ where $n = |V|$ is $O(n^3)$ is the Floyd algorithm. If we use an improved Bellman-Ford algorithm, the complexity will be $O(V^2 \log V + VE)$.

For each new point p, Algorithm 3.1 would need to check every point in D_0, which requires $O(D_0)$ time. Since the size of D_0 changes from $|J|$ to $|D|$, the total time of the algorithm T is

$$T = O(\Sigma |D_0|) = \Sigma_{i=|J|}^{n} i = O(n^2)$$

This is because in most cases $|J|$ is much smaller than $|D|$. $\square$

Algorithm 3.1 is the simplest algorithm to understand. A better implementation of the algorithm involves a data structure in the algorithm. To complete this, a queue for storing un-assigned value points is needed. The more detailed steps of the algorithm are found in Algorithm 3.2. In Chap. 8, we provide more insightful algorithm design issues.

In the case study section, we present a pseudo-code algorithm for this basic algorithm.

Algorithm 3.2 (Algorithm Using Breadth-First-Search*). All vertices are stored in a linked adjacency list. To generate such a list, we need $O(|E|)$ time.

Step 1:   Test if for all p and p' in J, $d(p,p') \geq LD(p,p)$. If not, then there is no gradually varied function interpolation. Otherwise, $D_0 \leftarrow J$.

Step 2:   Let Q be a queue that stores all vertices, each of them with an adjacent vertex in J. The new points are called J_1. Repeat this step such that J_2 are the new points. Each point in J_2 has an adjacent point in J_1. Q will be arranged by J_1, followed by $J_2, \ldots$ etc.

Step 3:   The rest of the algorithm will be the same as Algorithm 3.2.

3.4.2 The Divide-and-Conquer Algorithm

A fast algorithm can usually be applied to gradually varied reconstruction. This is because most of the domain has the Jordan separation property: a closed curve (a path in discrete cases) can separate D into two components that are not connected. For instances, four-connectivity in Σ_2 and triangulated decomposition have such a property.

We can use the divide-and-conquer technique here in the algorithm design. Let's assume first that J is the boundary of D. See Fig. 3.2a. The illustration is shown in Fig. 3.2b to explain how we use this method. We first separate the points into two equal components by finding a middle, dividing line.

In each component, only $|D|/2$ points will be checked until the recursive algorithm fills in all the points in D [3].

For general cases (J may not be the boundary), the method is still the same: cut the domain into two equal components, filling in the dividing line first. This line will split D into two components. Recursively cut each component into two and the algorithm will reach optimum. The complexity will be $(n \log n)$, called log-linear time, if all $d(x,y)$ are known.

The domain with a connectivity does not maintain a strict Jordan property. Thinking of bandwidth, there would be a band that can stop communication (direct connection) between the two components. Using this algorithm can help us find such a band. For Σ_2 with indirect adjacency, we can use two adjacent parallel lines to cut the domain. The algorithm would still be log-linear.

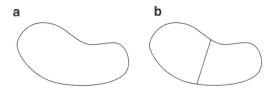

Fig. 3.2 Divide-and-Conquer Algorithm for Boundary known (**a**) Boundary of a Jordan domain; (**b**) Dividing into two components using a line

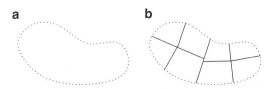

Fig. 3.3 Divide-and-Conquer Algorithm for Arbitrary Points (**a**) The region; (**b**) Recursively dividing into components

Algorithm 3.3 (Algorithm Using the Divide-and-Conquer technique*). [4, 9] All vertices are stored in a linked adjacency list. To generate such a list, we need $O(|E|)$ time.

Step 1:   Test if for all p and p' in J, $d(p,p') \geq LD(p,p)$. If not then there is no gradually varied function interpolation. $D_0 \leftarrow J$.

Step 2:   Find the dividing (middle) line that can split the domain D into two almost equal components. Then, fill the points on this dividing line.

Step 3:   For each component, repeat Step 2 until all points in D are filled.

Theorem 3.3. *The divide-and-conquer algorithm (Algorithm 3.3) for gradually varied functions has time complexity $O(n\log n)$ if we know $d(x,y)$ for all x,y in D.*

Proof. The key to the algorithm is cutting the domain in half. Roughly speaking, we have a recurring equation for this problem

$$T(n) = 2T\left(\frac{n}{2}\right) + B(n),$$

where $B(n)$ indicates the time for fitting the (middle) dividing line. Let's prove the case while D is a square, where $B(n) = O(n)$. We have

$$T(n) = 2T\left(\frac{n}{2}\right) + Cn$$

where C is a constant. According to the Master's theorem in [12],

$$T(n) = n\log(n)$$

□

The idea of this algorithm was first presented by Chen in 1989 [3] and a complete proof was presented in [4]. For some special domains divide-and-conquer algorithms based on the Jordan property were presented in [4]. The limited band Jordan case was discussed by Chen-Adjei in 2004 [11].

3.4.2.1 The Algorithm in Euclidean Space

Discrete distance in Euclidean space can be calculated in constant time in either four-connectivity or eight-connectivity. If we choose a length for the grids, Δ_x, and Δ_y, then the following are true:

$$d_4(p_1, p_2) = \frac{|(x_1 - x_2)|}{\Delta_x} + \frac{|(y_1 - y_2)|}{\Delta_y}. \tag{3.1}$$

$$d_8(p_1, p_2) = \max\{\frac{|(x_1 - x_2)|}{\Delta_x}, \frac{|(y_1 - y_2)|}{\Delta_y}\}. \tag{3.2}$$

The calculation of $d(x, y)$ for all x, y in a convex D is $O(n^2)$ when D is convex. If D is not convex, the distance does not follow the formula above. For 3D or higher dimension cases, the method is similar to that of convex domains. So the calculation of the distance between a pair of points is $O(|D|^2)$. Thus, in this case, there is no need for a Floyd algorithm (for Floyd algorithms, see [12]). We have,

Theorem 3.4. *Let D be a region in Euclidean space. The divide-and-conquer algorithm, Algorithm 3.3, has time complexity $O(n \log n)$.*

3.5 Case Study: An Example of Gradually Varied Fitting

Let us first look at the example below: Assume the array shown below, we use four-adjacency in this example. Δ is the position of the next value to be determined.

−	2	−	−	−		−	2	−	−	−		−	2	−	−	−		−	2	−	−	−		−	2	−	−	−
−	Δ	−	−	5		−	2	Δ	−	5		−	2	2	−	5		−	2	3	−	5		−	2	3	−	5
−	−	−	5	−		−	−	−	5	−		−	−	−	5	−		−	−	−	5	−		−	−	4	5	−
3	−	−	4	−		3	−	−	4	−		3	−	−	4	−		3	−	−	4	−		3	−	−	4	−
−	4	5	5	−		−	4	5	5	−		−	4	5	5	−		−	4	5	5	−		−	4	5	5	−

The first Δ will be replaced by 2 (its neighbor's value), then check if 2 satisfies the gradual variation condition with all other values in the array. It is true. So we keep the value of 2. After that, we put 2 at the second Δ, but it is not satisfying the gradual variation condition, thus, we change it to be 3. And so on. We give pseudo-code here that can easily be translated into C++ code [7].

Procedure A: Given a digital manifold D and its subset J and given that $f_J : J \to \{A_1, \ldots, A_m\}$, where $A_1 < \ldots < A_m$, this procedure can be used to obtain a gradually

varied function $f : D \to \{A_1, \ldots, A_m\}$ or conclude that a gradually varied function f
for f_J does not exist.

```
BEGIN
        FOR (every pair p, p' in D) DO
            compute d(p, p') by Floyd's algorithm [6];
        FOR (every pair p, p' in J) DO
            IF (d(p, p') < d(f_J(p), f_J(p'))) THEN {there is no f and halt;}
        D_0 := J;
        FOR (every r ∈ D_0) DO { f(r) := f_J(r); }
5       IF (D_0 = D) THEN
            output f and halt;
        ELSE
            BEGIN
                choose x in D − D_0 such that x has an adjacent point
                r ∈ D_0. (without loss of generality assume f(r) = A_i);
            END
        f(x) := A_i;
        FOR (every p ∈ D_0) DO
            BEGIN
                IF( d(x, p) < d(f(x), f(p)) ) THEN
                    IF ( f(x) > f(p) ) THEN ( f(x) := A_{i−1} ) ELSE ( f(x) := A_{i+1} );
            END
        D_0 := D_0 ∪ {x};
        GOTO 5;
END
```

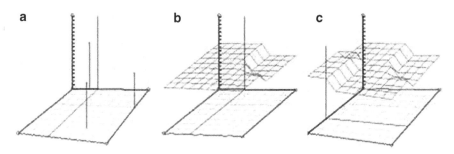

Fig. 3.4 Gradually varied interpolation. (**a**) Data samples, (**b**) Surface reconstruction, (**c**) Add a sample point and update results

3.6 Properties of Gradual Variation*

Gradually varied functions have many interesting properties. An envelope theorem, a uniqueness theorem, and an extension theorem preserving the norm were obtained in [7]. These results can be used to design specific gradually varied surfaces.

Without loss of generality, we use Σ_1, a chain space, to represent $\{1,2,\cdots,n\}$ or $\{A_1,A_2,\cdots,A_n\}$. In Chap. 8, we will extend the range space to more general cases.

The envelope theorem is used to set a boundary or margin for a gradually varied surface when the guiding points are given.

Theorem 3.5 (Chen 1994). *If J is a non-empty subset of D and a mapping $f_J : J \rightarrow \Sigma_1$ has a gradually varying extension on D, then there are two gradually varied functions S_1 and S_2 on D, such that every gradually varied function S is between S_1 and S_2. In other words, for any $x \in D$, $S_1(x) \geq S(x) \geq S_2(x)$.*

Suppose that p and p' in J satisfy $d(p,p) = d(f(p),f(p'))$, then the valuations of each point in the shortest path between p and p' are unique. This is called a non-abundant path [4]. A uniqueness theorem is to observe that, in a given situation, there is only one possible gradually varied surface interpolation.

Theorem 3.6 (Chen, 1992, 1994). *Let J be a non-empty subset of D and a mapping $f_J : J \rightarrow \Sigma_1$ has a gradually varying extension on D, then there is only one unique gradually varied function on D if and only if D is covered by a non-abundant number of shortest paths whose two end points are in J.*

The proof can be found in [10]. A related topic is to count how many gradually varied surfaces there are if guiding points are given. This problem has not yet been solved. We do not even know whether it is an NP-hard problem [4, 12]. We did some preliminary research of this counting problem in [4].

Let f_D be a gradually varied functional on D, the length of the longest non-abundant path is called the norm of f_D, which is denoted by $||f_D||$ in this book. $||f_D||$ has some strongly geometric meanings.

Theorem 3.7 (Chen 1994). *Let f_J be a function defined on a non-empty subset J of D. If for some $n > 0$, then:*

1. For all p and p' in J, $d(p,p') \geq d(f_J(p),f_J(p'))$;
2. If $d(f_J(p),f_J(p')) = k \cdot n + i$, $0 \leq i < n$ then

   (a) $d(p,p') \geq k \cdot n + (k-1)$, $i = 0$,
   (b) $d(p,p') \geq k \cdot n + k + i$, $i \neq 0$,

then there exists a gradually varied extension f_D of f_J with $||f_D|| \leq n$.

If there is no gradually varied interpolation for J, one can still perform a gradually varied approximation practically according to the following theorem.

Without loss of generality, assume $A_i = i$ for convenience. Let us denote

1. $F_i(p) = \{f_J(p) \pm t \mid t = 0,1,\ldots,i\}$,
2. $F_i(p)/F_i(q) = \{x \mid x \in F_i(p) \text{ and there exists } y \in F_i(q) \text{ such that } d(p,q) \geq d(x,y)\}$,
3. $F_i^k(p) = \cap_{q \in J} F_i^{k-1}(p)/F_i^{k-1}(q)$,

where $F_i^0(p) = F_i(p)$.

Theorem 3.8 (Chen, 1992, 1994). *(1) For every p and k, $F_i^k(p) \subset F_i^{k-1}(p)$. (2) If there exists a k for all $p \in J$, $F_i^k(p) = F_i^{k+1}(p)$, then for all $N > k$, $F_i^k(p) = F_i^N(p)$.*

Theorem 3.9 (Chen, 1992, 1994). *If $\exists k \forall p \in J(F_i^{k+1}(p) = F_i^k(p) \neq \emptyset)$, then for every $p \in J$, let $g_J(p) = \inf\{F_i^k(p)\}$. Then, g_J satisfies that for any $p, q \in J$, $d(p,q) \geq d(g(p), g(q))$, and $\forall_{p \in J}(d(g(p), f(p)) \leq i)$.*

The detailed proofs of the above theorems can be found in [10].

3.7 Mathematical Foundation of Gradually Varied Functions*

Constructive Mathematics is based on the constructability of mathematical concepts. It is the most reliable mathematics compared to formalistic mathematics. In this section, we present a foundation for gradually varied functions based on constructive mathematics [2, 10].

Constructive functions are defined based on constructive analysis, which is in some extent similar to algorithmic constructible functions. The difference is that the former is based on separable space, especially constructive-compact-metric-space, and the latter is based on the Turing machine. Since this book is about finite discrete methods. It is important for us to introduce the concept of constructive functions and their approximations.

In mathematics, a (topological) space X is said to be separable if it contains a countable dense subset C, meaning that for every point x in X and a small number r, there will be a point c in C such that $|x - c| \leq r$. For example, the real number set R is separable because it contains the rational number set, which is dense in R.

Therefore, in numerical analysis and computer science, it is very important to know that people now can only deal with finite sets, or a subset of rational numbers.

A function that is constructive usually means that there is a procedure to obtain approximate values that converge to the value it is actually supposed to be. To make this discussion more concise, we only need to discuss one lemma in constructive compact space.

The meaning of compactness is sometimes difficult to explain. It is used to find a convergence point for an infinitive sequence. It states that if there is an open cover for space X, then there much be a finite subset of the open cover that covers X.

If there is a constructive way to find such a finite cover, it can be seen that there will be an infinite number of points in an open set. So when this open set is small, we can find a convergence point "algorithmically." For every arbitrary open set collection

$$\{B_\alpha\}_{\alpha \in O}, \tag{3.3}$$

where O is a (constructive, or programmable) set, if

$$X = \bigcup_{\alpha \in O} B_\alpha, \tag{3.4}$$

then there will be a finite subset of O, $J \subset O$ such that

$$X = \bigcup_{i \in J} B_i. \tag{3.5}$$

Such X is called a compact set.

In general, the model for constructive analysis is larger than that of the Turing machine. It accepts the concept of two equal real numbers that are generally undecidable using the Turing machine.

The intuitive meaning of separable space is that one can insert a nonempty open set between any two points x and y. To define this, we can use two neighborhoods around x and y, namely $N(x, r_1)$ and $N(y, r_2)$, respectively. (r_1 and r_2 are two radiuses.) We require $N(x, r_1) \cap N(y, r_2) =$. Such a separable space is also called a Hausdorff space. It is not obvious that this definition can be realized by a program.

In the foundation of mathematics [2], constructive mathematics is more close to human's intuition. Basically, a mathematical object is constructible if there is a routine to generate it. The routine here contains more information then an algorithm. It is much "safer" than formalism mathematics that is built on axioms and proof theory. For instance, to decide that mathematical system is consistency is impossible within the system in general. Therefore, the constructive method for gradually varied functions is trustable.

Suppose M is a constructive-compact-metric-space described by Bishop and Bridges [2]. Then there exists a routine which generates a sequence of a finite set $M_1, \ldots, M_n, \ldots$, such that M_n is a 2^{-n} net of M. Then M_n is a digital manifold, and the adjacent set of $x \in M$ is

$$N(x) = \{y | y \in M_n \& w(x, y) < 2^{-n}, \text{ where } w \text{ is the metric of } M\}.$$

Let f be a continuous functional on M (f is also a uniform continuous functional [2]). Therefore, for any l, there exists n such that: if $w(x, y) < 2^{-n}$, then $|f(x) - f(y)| < 2^{-l}$.

Now, we construct a functional f_n on M_n as follows. Let $A(s) = s \cdot 2^{-l}$, and set $f_n(x) = A(s)$, where $s = \min\{t | f(x) \leq A(t) \& x \in M_n\}$. Then f_n is a 2^{-n}-uniform approximation of f on M_n.

Lemma 3.1. $f_n : M_n \to \{A(s)\}$ is gradually varied on M_n.

Proof. If x and y are adjacent in M_n, then $w(x, y) < 2^{-n}$, and $|f(x) - f(y)| < 2^{-l}$. If $f_n(x) \leq f_n(y)$, then $f(x) \leq f_n(x) \leq f_n(y)$. Thus,

$$0 \leq f_n(y) - f_n(x) \leq f_n(y) - f(x) \leq f_n(y) - f(y) + f(y) - f(x) < 2 \cdot 2^{-l},$$

so $|f_n(x) - f_n(y)| < 2 \cdot 2^{-l}$. Similarly, in case of $f_n(x) > f_n(y)$, we can also obtain $|f_n(x) - f_n(y)| < 2 \cdot 2^{-l}$. That is, assume $f_n(x) = A(s)$ and $f_n(y) = A(t)$, then we must have $|t - s| < 2$. Hence, f_n is gradually varied. $\diamondsuit$

(M_n, f_n) is said to be an (n, l)-uniform approximation of (M, f). The process of generating $\{(M(n), f_n), n \in I\}$ is called gradually varied approximating. In calculus, Weierstrass's approximation theorem tells us that for any $\varepsilon > 0$ there exists a polynomial f_p such that f_p is an ε-uniform approximation of f. That is to say, the gradually varied function is reasonably proposed.

In addition to the concept of gradually varied functions (or functionals), n is a function of l. If there is an $m > 0$, such that

$$\lim_{l \to \infty} \frac{n(l)}{l^m} = O(1), \qquad (3.6)$$

then m is called the order of the gradually varied function.

For an arbitrary M, f, and $\varepsilon > 0$, there exists an m so that we can generate an mth-order gradually varied function that is an ε-uniform approximation of f.

3.8 Remark

The gradually varied surface was originally designed for digital surface fitting without making any assumptions about continuous functions such as splines or Bezier polynomials in geophysical data processing [3–5]. A gradually varied surface can be viewed as a discrete function f from Σ_2 to $\{1, 2, \ldots, n\}$ satisfying $|f(a) - f(b)| \le 1$ if a and b are adjacent in Σ_2. This concept is called "discretely continuous" by Rosenfeld [15] and "roughly continuous" by Pawlak [14]. A gradually varied function can be represented by λ-connectedness, which will be introduced in Chap. 12.

For Theorem 3.1, Rosenfeld in 1986 attempted to find a connection between two connected components that was related to this theorem [15]. The Lemma 3.1 in Section 3.7 was reviewed by D. Bridges, one of the authors of [2].

References

1. Aho A, Hopcroft JE, JD Ullman (1974) The design and analysis of computer algorithms. Addison-Wesley, Reading
2. Bishop E, Bridges D (1985) Constructive analysis. Springer, Berlin/New York
3. Chen L (1990) The necessary and sufficient condition and the efficient algorithms for gradually varied fill. Chinese Sci Bull 35(10):870–873. Abstracts of SIAM conference on geometric design, Tempe, AZ, 1989
4. Chen L (1991) The properties and the algorithms for gradually varied fill. Chinese J Comput 14:3
5. Chen L (1992) Random gradually varied surface fitting. Chinese Sci Bull 37. Abstracts of second SIAM conference on geometric design, Tempe, AZ, 1991 37(16):1325–1329
6. Chen L (1991) Gradually varied surfaces on digital manifold. In: Abstracts of second international conference on industrial and applied mathematics, Washington, DC, 1991
7. Chen L (1994) Gradually varied surface and its optimal uniform approximation. In: *IS&T* SPIE Symposium on Electronic Imaging, SPIE Proceedings, vol 2182. (L. Chen, Gradually varied surfaces and gradually varied functions, in Chinese, 1990; in English 2005 CITR-TR 156, U of Auckland. Has cited by IEEE Trans in PAMI and other publications.)

8. Chen L (1997) Generalized discrete object tracking algorithms and implementations. Vision geometry VI, Proceedings SPIE, vol 3168, San Diego
9. Chen L (1999) Note on the discrete jordan curve theorem. Vision geometry VIII, Proceedings SPIE, vol 3811, 1999
10. Chen L(2004) Discrete surfaces and manifolds: a theory of digital-discrete geometry and topology, Scientific and Practical Computing, Rockville, MD
11. Chen L, Adjei O (2004) Lambda-Connected segmentation and fitting. In: Proceedings of IEEE international conference on systems man and cybernetics, vol 4, pp 3500–3506, Hague, Netherlands
12. Cormen TH, Leiserson CE, Rivest RL (1993) Introduction to algorithms. MIT, New York
13. Minkowsky H (1896) Geometrie der Zahlen. Teubner, Leipzig
14. Pawlak Z (1999) Rough sets, rough functions and rough calculus. In: Pal SK, Skowron A (ed) Rough fuzzy hybridization. Springer, Singapore, pp 99–109
15. Rosenfeld A (1986) "Continuous" functions on digital pictures. Pattern Recognit Lett 4:177–184
16. Rosenfeld A, Kak A (1982) Digital picture processing, vols I and II. Academic, Orlando

Chapter 4
Gradually Varied Extensions

Abstract In this chapter, we generalize the concept of gradually varied functions to gradually varied mappings. In Chap. 3, we mainly discussed the function from a discrete space to $\{1, 2, \cdots, n\}$ or $\{A_1, A_2, \cdots, A_n\}$. This chapter focuses on gradually varied functions from a discrete space to another discrete space. For instance, a digital surface is defined as a mapping f from an n-dimensional digital manifold D to an m-dimensional grid space Σ_m. A discrete surface is said to be gradually varied if two points in D, p and q, are adjacent, implying that $f(p)$ and $f(q)$ are adjacent in Σ_m. In this Chapter, a basic extension theorem will be proven and some counter examples will be presented.

4.1 Basic Gradually Varied Mapping and Immersion

In mathematics, a narrow definition of function is a mapping from a set to the real number set R. This chapter generalizes the concept of gradually varied functions in Chap. 3 to more general gradually varied mappings. In Chap. 3, we defined:

Let $G = (V, E)$ be a graph and $\{A_1, A_2, \cdots, A_n\}$ be a set of real numbers with $A_1 < A_2 < \cdots < A_n$. A function $f : G \to \{A_1, A_2, \cdots, A_n\}$ is gradually varied if for any adjacent pair p, q in G, $f(p) = A_i$ implies $f(q) = A_{i-1}$, A_i, or A_{i+1} [4, 6–8, 10, 17].

We know that $A_1 < A_2 < \cdots < A_n$ form a chain or a path [12, 15]. If we extend this range to a more general space, the gradually varied function will then be called a gradually varied mapping in order to separate it from the more popular definitions of these terms.

Here, we focus on gradually varied extension, in other words the existence of a mapping from one (discrete) manifold to another. In this chapter, some extension theorems will be proven and some counter examples will be presented.

In order to make the definition of gradually varied mapping suitable for standard mathematical terminology, in this chapter we use the concept of immersion in the same way as the concept of gradually varied mapping. In addition, the concept of discrete manifolds is defined as the discretization of smooth manifolds or simply mean graphs. A more precise definition for discrete manifolds is presented in Chap. 8 or in [10, 12, 13].

Definition 4.1 ([11, 12]). Let D_1 and D_2 be two discrete manifolds and $f : D_1 \to D_2$ be a mapping. f is said to be an immersion from D_1 to D_2, or a gradually varied operator, if x and y are adjacent in D_1, implying that $f(x) = f(y)$ or $f(x)$ and $f(y)$ are adjacent in D_2.

If $D_2 = \Sigma_m$, then f is called a gradually varied surface . A digital surface is defined as a mapping f from an n-dimensional digital manifold D to an m-dimensional grid space Σ_m. A discrete surface is said to be gradually varied if two points in D, p and q, are adjacent, implying that $f(p)$ and $f(q)$ are adjacent in Σ_m.

An immersion f is said to be an embedding if f is a one-to-one mapping. In fact, D_1 and D_2 can be two simple graphs in the above definition. In this case, we know a famous NP-complete problem [2, 14], the subgraph isomorphism problem [14], is related to the gradually varied operator.

Lemma 4.1. *There is no polynomial time algorithm to decide if a graph D can be embedded within another graph D' unless $P = NP$.*

However, if the number of adjacent points to each point in D and D' are not greater than a constant c and $|D| = |D'|$, then there exists a polynomial time algorithm to decide whether $|D|$ and $|D'|$ are isomorphic [14]. On the other hand, the immersion problem is trivial (i.e. the existence of immersion) because we can always let the image of all points in D be a certain point in D'.

There is another related NP-complete problem. An immersion is said to be a morphism if for $a' = f(a)$, where b' is an adjacent point to a', then there is a $b \in f^{-1}(b')$ such that a, b are adjacent. Deciding whether or not an immersion is a morphism is NP-hard.

There are many interesting problems in algorithm design and computational complexity. However, in this book, we focus on how to obtain the entire mapping using the property of immersion on a given part of a mapping.

4.2 The Main Theorem of Gradually Varied Extension

Mathematically, the main problem of the chapter is as follows: Let D and D' be two discrete manifolds. Assume J is a subset of D. If $f_J : J \to D'$ is known, then there is an extension of f_J, f_D, such that $f_D : D \to D'$ is gradually varied, where an extension is defined as $f_J(a) = f_D(a)$ if $a \in J$ [5].

It is easy to see that if f is a gradually varied mapping, then

$$\forall_{p,q \in D} d_D(p,q) \geq d_{D'}(f(p),f(q)), \tag{4.1}$$

where $d_D(x,y)$ is the length of the shortest path between x and y in D, and $d_{D'}(u,v)$ is the length of the shortest path between u and v in D'.

In Chap. 3, we discussed a case when D' is an ordered set $\{1,2,3,\cdots,n\}$. We have presented a related theorem, Theorem 3.1 in Chap. 3 that states the necessary and sufficient conditions to the existence of gradually varied functions for the given sample points in D.

Since an ordered set can be represented as a chain, Theorem 3.1 can be presented as follows: Let J be a subset of D, and D' be a chain. If f_J satisfies

$$\forall_{p,q \in J} d_D(p,q) \geq d_{D'}(f(p),f(q)), \tag{4.2}$$

then there is a gradually varied extension f_D of f_J. We also discussed fast algorithms that can be directly used for gradually varied extension in Chap. 3.

We can see that Theorem 3.1 is an ideal case. However, not every D,D' can preserve such a property, i.e. the existence of gradually varied extension. For example, consider D and D' as shown in Fig. 4.1a, b, respectively. If $J = \{a,b,c,d\}$ and f_J is indicated in Fig. 4.1b, then we can see that $< J, f_J >$ satisfies the condition in Theorem 3.1. But for point x, there is no $f(x)$ that can be selected to satisfy the condition of gradual variation.

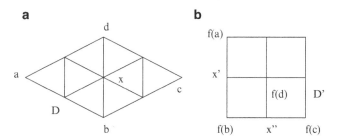

Fig. 4.1 Example that shows no gradually varied extension

In order to do more profound research and make the terminology simpler, we propose a new concept as follows:

Definition 4.2. Let J be a subset of D and f_J be a mapping $f_J : J \to D'$, which satisfies:

$$\forall_{p,q \in J} [d(p,q) \geq d(f_J(p),f_J(q))]. \tag{4.3}$$

If there exists an extension f of f_J such that $f : D \to D'$ is a gradually varied mapping, then we say $< J, f_J >$ is immersion-extendable. If every $< J, f_J >$ satisfying (4.3) is immersion-extendable, then we say that D can normally immerse into D'.

Mathematically, it might be very difficult to decide whether or not a given graph D and function f on $J \subset D$ can immerse into another D'. This problem may be NP-hard [14]. However, in terms of data fitting, pattern recognition, and computer vision, we can find several important and useful digital manifolds in which normal immersion and general gradually varied surfaces are applicable. In other words, given some guiding points and their (mapping) values in such a discrete manifold, we can find a gradually varied surface based on the guiding points.

The m-dimensional grid space Σ_m is the set of all points in m-dimensional Euclidean space with integer coordinates. There are many kinds of adjacency, and we present two of them in Chap. 2, which are most often used in discrete geometry.

Recall the definition, two points $p = (p_1, \dots, p_m)$ and $q = (q_1, \dots, q_m)$ are directly adjacent if and only if $\sum_{i=1}^{m}(|p_i - q_i|) = 1$. Such a digital space is called a directly adjacent space and is denoted by d_Σ_m. Second, two points $p = (p_1, \dots, p_m)$ and $q = (q_1, \dots, q_m)$ are indirectly adjacent if and only if $Max_{i=1}^{m}(|p_i - q_i|) = 1$. Such a digital space is called an indirectly adjacent space and is denoted by i_Σ_m.

In addition, the n-simplicial decomposition space is denoted by Δ_n.

We know Theorem 3.1 states that any digital manifold can normally immerse to Σ_1 or a chain. We can show a more general theorem below [5, 9]:

Theorem 4.1. *Any graph D (or digital manifold) can normally immerse an arbitrary tree T.*

Proof. Suppose that $J \neq \emptyset$ is a subgraph of D, and a mapping $f_J : J \rightarrow T$ satisfies (4.3), where T is a tree. We will prove that there exists a gradually varied mapping $f : D \rightarrow T$ with $f_J(p) = f(p)$, $p \in J$.

The proof of the theorem is a constructive proof. For every $p \in J$, let $f(p) = f_J(p)$, and let $f(p) = \theta$ if $p \in D - J$. Denote

$$D_0 = \{p | f(p) \neq \theta \& p \in D\}. \tag{4.4}$$

For any p and p' in D_0, we have $d(p, p') \geq d(f(p), f(p'))$.

If $D_0 \neq D$, we can find a point x in $D - D_0$ such that x has an adjacent $r \in D_0$, then we have two cases:

Case 1. If $d(r, p) > d(f(r), f(p))$ for every $p \in D_0$ with $p \neq r$, then let $f(x) = f(r)$. Since $d(r, x) = 1$ is not less than $d(f(x), f(r)) = 0$ and

$$d(x, p) \geq d(r, p) - 1 > d(f(r), f(p)) - 1,$$

$d(x, p) \geq d(f(r), f(p)) \geq d(f(x), f(p))$. Let $D_0 \leftarrow D_0 \cup \{x\}$ and return to repeat.

Case 2. If there is a point $p \in D_0$ such that $d(r, p) = d(f(r), f(p))$, then let

$$R_1 = \{p | d(r, p) = d(f(r), f(p)), p \in D_0\}. \tag{4.5}$$

According to (4.5), there are only two instances:

(2.1) If for every p in R_1, $d(r,p) < d(r,x) + d(x,p)$, then $d(r,p) < 1 + d(x,p)$, so $d(x,p) \geq d(r,p)$. Let $f(x) = f(r)$. If $p \in R_1$, then $d(x,p) \geq d(r,p) \geq d(f(r),f(p)) \geq d(f(x),f(p))$; else, i.e., p in $D_0 - R_1$, since $d(r,p) > d(f(r),f(p))$ then

$$d(x,p) \geq d(r,p) - 1 \geq d(f(r),f(p)) \geq d(f(x),f(p)).$$

Let $D_0 \leftarrow D_0 \cup \{x\}$ and return to repeat.

(2.2) If there is some p in R_1 such that

$$d(r,p) = d(r,x) + d(x,p) = 1 + d(x,p),$$

then x is obviously the shortest path between r and p. In the following, we will fix the point p. For every $q \in D_0$ with $q \neq r$ and $q \neq p$, we have $d(p,q) \leq d(p,x) + d(x,q)$. Let

$$m(x) = \{q | d(x,q) \leq d(q,r), q \in D_0\}. \tag{4.6}$$

There are only two cases to discuss:

(2.2.1) If $q \in m(x)$, then $d(p,q) \leq d(p,x) + d(x,q) \leq d(r,p) - 1 + d(r,q)$. Also, since p belongs to R_1,

$$d(p,q) \leq d(r,p) + d(r,q) - 1, \tag{4.7}$$

and

$$d(p,q) \leq d(f(r),f(p)) + d(r,q) - 1. \tag{4.8}$$

Three cases can be derived by the above Eq. (4.8):

(2.2.1.1) If $d(r,q) \geq d(f(r),f(q)) + 2$, then $d(x,q) \geq d(r,q) - 1 \geq d(f(r),f(q)) + 1$. Therefore,

$$d(f(x),f(r)) \leq 1 \text{ implies } d(x,q) \geq d(f(x),f(q)). \text{ Just let } f(x) = f(r).$$

(2.2.1.2) If $d(r,q) = d(f(r),f(q))$, then by using (4.8) and $p \in R_1$,

$$d(p,q) \leq d(f(r),f(q)) + d(f(r),f(p)) - 1. \tag{4.9}$$

Because T is a tree, T has one and only one path between $f(p)$ and $f(q)$, which must be the shortest path. If this path contains the point $f(r) \in T$, then

$$d(f(p),f(q)) = d(f(r),f(q)) + d(f(r),f(p)) \tag{4.10}$$

and by using (4.9), we can obtain

$$d(p,q) \leq d(f(p),f(q)) - 1.$$

A contradiction occurs from the assumption that p and q are in D_0. Hence, this path is not through $f(r)$.

One can view $D' = T$ as the tree with $f(r)$ as the root. When the path does not go through $f(r)$, $f(p)$ and $f(q)$ must be in the same proper subtree of T. There must be an α that is adjacent to $f(r)$ such that

$$d(\alpha, f(p)) = d(f(r), f(p)) - 1$$
$$d(\alpha, f(q)) = d(f(r), f(q)) - 1.$$

Let $f(x) = \alpha$. We can easily get $d(x, p) \geq d(f(x), f(p))$. And for all q, we have $d(x, q) \geq d(f(x), f(q))$ [5].

(2.2.1.3) If $d(r, q) = d(f(r), f(q)) + 1$, then the following applies:

  (i) If $d(x, q) \geq d(r, q)$, then $d(x, q) \geq d(f(r), f(q)) + 1$. (This case is similar to 2.2.1.1).

  (ii) If $d(x, q) = d(r, q) - 1$ and according to the fixed p with $d(r, p) = 1 + d(x, p)$ and $d(r, p) = d(f(r), f(p))$, we have

$$d(p, q) \leq d(p, x) + d(x, q) \leq d(r, q) + d(r, p) - 2,$$

so $$d(p, q) \leq d(f(r), f(q)) + 1 + d(f(r), f(p)) - 2.$$
(This is similar to 2.2.1.2).

(2.2.2) If $q \in D_0 - m(x)$, i.e., $d(x, q) \geq d(r, q) + 1$, then

$$d(x, q) \geq d(f(r), f(q)) + 1.$$

(This is similar to 2.2.1.1). Let $f(x) = f(r)$. (*We have completed the mathematical proof here. The following discussion relates to algorithm design.*) To summarize, if there is a point p in R_1, such that

$$d(r, p) = 1 + d(x, p) = d(r, x) + d(x, p),$$

then for any q in $D_0 - \{r, p\}$, it must be included in M_1 or M_2, where M_1 is

$$M_1 = \{q \mid d(x, q) \geq d(f(r), f(q)) + 1, q \in D_0 - \{r, p\}\}.$$

We can choose $f(x) = f(r)$ or make $f(x)$ adjacent to $f(r)$ in the tree T.

$$M_2 = \{q \mid \text{ there exists a shortest path between } f(p) \text{ and } f(q)$$
$$\text{that does not go through } f(r), \ q \in D_0 - \{r, p\}\}$$

Let $F_2 = \{f(q) \mid q \in M_2\}$. We can see that point $f(r)$ is the root of the tree T. It is easier to see that every point in $F_2 \cup \{f(p)\}$ is included in a subtree of the tree. Thus there exists a node t in T, such that $d(f(r), t) = 1$ and

$$\forall_{s \in F_2 \cup \{f(p)\}} d(f(r), s)) = d(t, s) + 1.$$

Therefore, let $f(x) = t$. In fact, if the path between $f(r)$ and $f(p)$ is $f(r), s_1, \ldots, s_{l-1}, f(p)$, then $t = s_1$. We have now proven Theorem 4.1. $\square$

Now we can immediately get,

Corollary 4.1. *Any graph/digital manifold can normally immerse into an arbitrary forest.*

The proof of Theorem 4.1 is relatively complicated but still elementary. In 2006, Agnarsson and Chen obtained a general theorem involving the Helly property : Let S be a set. A family of subsets has the Helly property if the intersection of every two subsets is not empty, meaning the intersection of all subsets contains at least one element [1].

Theorem 4.2. *For a reflexive graph G the following are equivalent: (1) G has the extension property (2) G is an absolute retract . (3) G has the Helly property.*

An alternate representation of the theorem is as follows: For a discrete manifold M the following are equivalent: (1) Any discrete manifold can normally immerse M. (2) Reflexivized M is an absolute retract. (3) M has the Helly property.

We do not give a proof of this theorem here since more background knowledge in graph theory is needed for the proof.

4.3 Gradually Varied Extensions for Indirect Connectivity of Σ_m

We now define i_Σ_m as the space Σ_m under indirect adjacency. That is, two points $p = (x_1, \cdots, x_m)$ and $q = (y_1, \cdots, y_m)$ in Σ_m are said to be adjacent if

$$d_I(p,q) = \max_{i=1}^{m} |x_i - y_i| = 1, \qquad (4.11)$$

d_Σ_m will be defined as the space Σ_m under direct adjacency: two points $p = (x_1, \cdots, x_m)$ and $q = (y_1, \cdots, y_m)$ in Σ_m are said to be directly adjacent if

$$d_D(p,q) \sum_{i=1}^{m} |x_i - y_i| = 1 \qquad (4.12)$$

Theorem 4.3 ([5, 9]). *Any graph/digital manifold D can normally immerse into i_Σ_m.*

Proof. Let D be a digital manifold. Suppose J is a subset of D, $J \neq \emptyset$, and the mapping $f_J : J \to i\_\Sigma_m$ satisfies (4.3). We can denote

$$f_J(x) = (f_J^{(1)}(x), f_J^{(2)}(x), \ldots, f_J^{(m)}(x)),$$

and we have, for any two points p and q,

$$d(p,q) = d_I(p,q) \geq d(f_J(p), f_J(q)).$$

Since

$$d(f_J(p), f_J(q)) = Max\{d(f_J^{(k)}(p), f_J^{(k)}(q)) | k = 1, \ldots, m\},$$

then

$$d(p,q) \geq d(f_J^{(k)}(p), f_J^{(k)}(q)), k = 1, \ldots, m.$$

Let $f_J^{(k)} : J \to i\_\Sigma_m$, where

$$f_J^{(k)}(x) = (0, \ldots, 0, f_J^{(k)}(x), 0, \ldots, 0). \tag{4.13}$$

By Theorem 3.1, we have m gradually varied extensions as follows:

$$f^{(k)} : D \to i\_\Sigma_m, k = 1, \ldots, m. \tag{4.14}$$

Finally, we can combine the m gradually varied extensions into a unique mapping:

$$f(x) = \sum_{k=1}^{m} f^{(k)}(x),$$

preserving $f(p) = f_J(p)$, $p \in J$. Also, if p and q are adjacent in D, then by (4.11)

$$d(f(p), f(q)) = \max\{d(f[k](p), f[k](q)) | k = 1, \ldots, m\} \leq 1.$$

Hence, f is a gradually varied extension of f_J. $\square$

4.4 Some Counter Examples for Gradually Varied Extensions

Not every mapping has gradually varied extension. Some examples are as follows:

Theorem 4.4 ([5]).

(1) A d_Σ_m cannot normally immerse into itself where $m > 1$.
(2) Δ_n, where $n > 2$, cannot normally immerse into itself.
(3) d_Σ_m or i_Σ_m cannot normally immerse into Δ_n, where $n > 2$.
(4) Δ_n, where $n > 2$, cannot normally immerse into d_Σ_m.

Proof. We only need to list some counter-examples to prove this theorem.

(1) Assume the D, D', which are given in Fig. 4.2a, b. If $J = \{p,q,s,r\}$ and $f_J = \{p',q',s',r'\}$ with $f_J(a) = a'$, then we know for all $a,b \in J$, $d(a,b) \geq d(f_J(a), f_J(b))$. However, for $x \in D$, $f_J(x)$ can only be assigned as x', or x''. Neither x' nor x'' can satisfy the condition for gradual variation.

(2) For $< \Delta_n, \Delta_m >$, where $m,n > 2$, let the mapping f_J have domain $J = \{a,b,c\}$ and range $f_J = \{a',b',c'\}$ in Fig. 4.2c, d. We cannot find the image of $x \in D$ satisfying the condition of gradual variation.

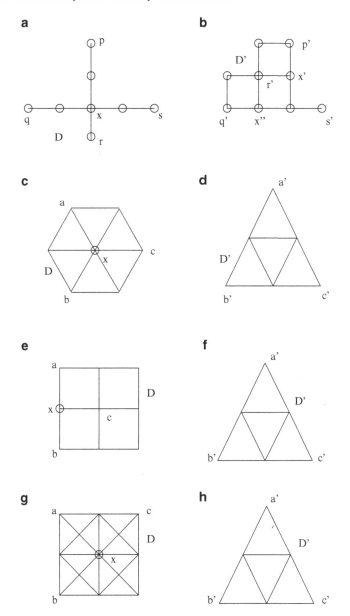

Fig. 4.2 Examples that show no gradually varied extension

(3) For $< d\_\Sigma_m, \Delta_n >$, the proof is similar to that of (2), with respect to Fig. 4.2e, f.

(4) For $< i\_\Sigma_m, \Delta_n >$, the proof is also similar to that of (2) with respect to Fig. 4.2g, h. □

The above results tell us that with d_Σ_m or Δ_n as the range-domain, there is basically no "continuous image."

A real world concrete example that supports the above results is the RGB-color table. Suppose we have domain $D = \Sigma_2$ and range $D' = \Delta_3$ with length C (See Fig. 4.2e). D' has three vertices R, G, and B at the three corners of D. R, G, and B represent "red," "green," and "blue." For any point x' in D', $d(x,R) + d(x,G) + d(x,B) = 2C$. So, if a point $x \in D$ and $f_J(x) = x'$, then we set the color $(r,g,b) = (C - d(x,R), C - d(x,G), C - d(x,B))$ to point x. We know the total color density of each point is the same, $C = r + g + b$. Theorem 4.4 tells us that if we wish to color selected points in D, it is not always possible to get a "continuous" looking color image using an equal density of colors.

The question that arises is: under what conditions does this become possible? We know that it is not only "for all $x, y \in J$, $d(x,y) \geq d(f_J(x), f_J(y))$." In practice, one color is always fixed as a base color and the other two colors are adjusted to get a continuous looking part of an image.

Definition 4.3. A (point-) connected component of a discrete manifold D is called a discrete sub-manifold S. A sub-manifold is said to be semi-convex if for any two points x, y in S, $d_S(x,y) = d_D(x,y)$.

The following is not difficult to prove:

Corollary 4.2. *If D can normally immerse into D', then any semi-convex sub-manifold of D can also normally immerse into D'.*

Corollary 4.3. *A complete graph K_n can normally immerse into any graph, and each graph can normally immerse into K_n.*

There are two computational complexity problems that are interested to solve:

(1) Is there a polynomial time algorithm that can decide if D is normally immersible into D' ?
(2) More specifically, for a given $< D, D', J, f_J >$, is there a polynomial algorithm that can obtain a gradually varied extension of f_J? We think the answer is no.

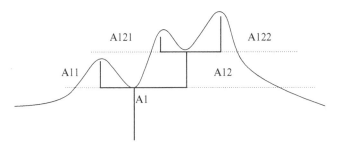

Fig. 4.3 A tree representation of discritization

4.5 Need of Gradually Varied Mapping and Extension

Assume that a continuous function f on $[0,2]$ is a parabolic on $[0,1]$ and an exponential on $[1,2]$.

$$f(x) := \begin{cases} (x-1)^2 + 1 & \text{if } 0 \leq x < 1 \\ e^{(x-1)} & 1 \leq x \leq 2 \end{cases}$$

It is obvious that if we use a total order of $A_1 < A_2 < \cdots < A_n$ to represent this combined function, we will not get an accurate level of discretization. This range space can be represented by a tree where each node on the tree will be assigned a value. For a different branch of the tree, the value could be the same but it would not carry the same meaning in terms of level of discretization.. Therefore, the chain structure of the image space (co-domain or range) is not enough.

A common method in image processing is pyramid representation based on the intensity of an image. For a time sequence it is well-known that a curve can be represented by a tree. It may be related to the Morse (height) function in topology. Using a tree instead of a chain for range space would solve the problem completely. In theory, this method was very well prepared in 1990s. Chen proved Theorem 4.1 stating that any graph can be normally immersed into a tree [7, 9]. This means that there is a gradually varied extension from a graph into a tree.

However, two A_i values that are the same but located in different branches of a tree means that they represent different elements. An example is shown below:

In Fig. 4.3, if $A_1 = \{B_0 < \cdots < B_k\}$, $A_{11} = \{C_0 < \cdots < C_r\}$, $A_{12} = \{D_0 < \cdots < D_t\}$, then $B_k = C_0 = D_0$.

In this example, B_1 may have the same numerical value as C_3, but it represents a different element in the tree. For a real world application, since the curve is not known, a classification technique will be used for grouping the sample points (again we can use decomposition technology). It may cause some uncertainty; however, for many real world problems this method is highly applicable and acceptable. The implementation for real problems sometimes may not only refer to algorithms in computer programming [1], but also mean constructible in human intuition [2].

4.6 Remark on General Discrete Surfaces and Extensions

In differential geometry in mathematics, a formal definition of surfaces is given through immersion mapping [16]. A digital surface is defined as a mapping f from an n-dimensional digital manifold D into an m-dimensional grid space Σ_m. A discrete surface is said to be gradually varied if two points in D, p and q, are adjacent, implying that $f(p)$ and $f(q)$ are adjacent in Σ_m.

We have proven the following constructive theorem: Let i_Σ_m be an indirectly adjacent grid space. Given a subset J of D and a mapping $f_J : J \to i\_\Sigma_m$, if the distance between any two points p and q in J is not less than the distance between $f_J(p)$ and $f_J(q)$ in i_Σ_m, then there exists an extension mapping f of f_J, such that the distance between any two points p and q in D is not less than the distance between $f(p)$ and $f(q)$ in i_Σ_m. That is to say, the guiding point set $(J, f(J))$ has a gradually varied surface fitting. In other words, any digital manifold (graph) can normally immerse into an arbitrary i_Σ_m. We also show that any digital manifold (graph) can normally immerse into an arbitrary tree T.

Another aspect of interest about related research for the functional extension problem was first studied by Valentine in 1945 [18]. We also presented a more general form of extension, Theorem 4.2, using Helly property in this chapter too [1].

References

1. Agnarsson G, Chen L (2006) On the extension of vertex maps to graph homomorphisms. Discrete Math 306(17):2021–2030
2. Aho A, Hopcroft JE, Ullman JD (1974) The design and analysis of computer algorithms. Addison-Wesley, Reading
3. Bishop E, Bridges D (1985) Constructive analysis. Springer, Berlin/New York
4. Chen L (1990) The necessary and sufficient condition and the efficient algorithms for gradually varied fill. Chinese Sci Bull 35:10. Abstracts of SIAM Conference on Geometric design, Tempe, AZ, 1989
5. Chen L (1990) Gradually varied surfaces and gradually varied functions, in Chinese; In English 2005 CITR-TR 156, University of Auckland (It has been cited by IEEE Trans in PAMI and other publications)
6. Chen L (1991) The properties and the algorithms for gradually varied fill. Chinese J Comput 14:3
7. Chen L (1991) Gradually varied surfaces on digital manifold. Abstracts of second international conference on industrial and applied mathematics, Washington, DC
8. Chen L (1992) Random gradually varied surface fitting. Chinese Sci Bull 37:16. Abstracts of second SIAM conference on geometric design, Tempe, AZ, 1991
9. Chen L (1994) Gradually varied surface and its optimal uniform approximation. *IS&T* SPIE symposium on electronic imaging, SPIE Proceedings, vol 2182, Sen Jose
10. Chen L (1997) Generalized discrete object tracking algorithms and implementations. Vision geometry VI, Proceedings SPIE, vol 3168, San Diego
11. Chen L (1999) Note on the discrete Jordan curve theorem. Vision Geometry VIII, Proceedings SPIE, vol 3811, Denver
12. Chen L (2004) Discrete surfaces and manifolds: a theory of digital-discrete geometry and topology. SP Computing
13. Chen L, Zhang J (1993) Digital manifolds: a Intuitive Definition and Some Properties. The Proceedings of the second ACM/SIGGRAPH symposium on solid modeling and applications, Montreal, pp. 459–460, 1993
14. Cormen TH, Leiserson CE, Rivest RL (1993) Introduction to algorithms. MIT, Cambridge, MA
15. Harary F (1972) Graph theory. Addison-Wesley, Reading
16. Mathematical Society of Japan (1993) Differential geometry of curves and surfaces. In: Encyclopedic of dictionary of mathematics, 2nd edn. MIT, pp 412–412, Cambridge, MA
17. Rosenfeld A (1986) 'Continuous' functions on digital pictures. Pattern Recognit Lett 4:177–184
18. Valentine FA (1945) A Lipschitz Condition Preserving Extension for a Vector Function. Am J Math 67(1):83–93

Chapter 5
Digital and Discrete Deformation

Abstract In science, changing one curve α into another curve β continuously is called deformation. To describe this action, we usually use a sequence of curves in a sketch: the beginning curve C_0 is the original curve α and the final curve C_1 indicates the targeting curve β. Therefore, deformation can be defined as a function $f^\alpha(t) = C_t$, where $t \in [0, 1]$. $f^\alpha(t)$ and $f^\alpha(t_0)$ are getting closer (infinitively) when $t \to t_0$. Such a concept has essential importance since it relates to the topological equivalence and effect on entire modern mathematics. It also has great deal of impact in 3D image processing, what we call morphing one 2D/3D picture into another. In this chapter, we introduce the basic method of digital deformation and homotopic equivalence. We also give a brief overview of the fundamental groups and homology groups for digital objects. (Note The material in this chapter is much different than that of other chapters because it contains some graduate level material in the mathematical field of topology. In this book, the author tries to explain some profound concepts in an elementary way, which may not always be successful meaning that it is not always appreciated by some others.)

5.1 Deformation and Discrete Deformation

A curve is a function. Changing this curve into another continuously is also a function. In this section, we introduce the concept of the function from one discrete object to another "continuously," usually called digital and discrete deformation in practice.

This method has direct applications to image processing and computer vision, such as morphing. It is often used to continuously turn one person's face into another through a sequence of images.

L.M. Chen, *Digital Functions and Data Reconstruction: Digital-Discrete Methods*, 51
DOI 10.1007/978-1-4614-5638-4_5, © Springer Science+Business Media, LLC 2013

5.1.1 Curves and Digital Curves in Space

A continuous curve can be mathematically defined as a function $f : [a,b] \to R$, such as the function shown in Fig. 2.2. In topology, the actual values of a and b are not very significant, so we use $[0,1]$ for the line segment. We usually use $P(t) = (x(t), y(t))$ to represent a 2D curve, so a 2D curve will be defined as $P : [0,1] \to R^2$, where $P(t)$ is continuous. Thus, we could extend this definition to an n-dimensional case: $P : [0,1] \to R^n$. It is called the parametric form of curves.

In digital space, a 2D digital "curve" is a digital continuous function $f : \{1,2,\dots, m\} \to \Sigma_2$. In general, any digital curve (or path) in 2D can be represented by a parametric digital continuous function $P(t) = (x(t), y(t)), t \in \{1,2,\dots,m\}$. where x and y are digital continuous functions. This simulated definition has some ambiguity in digital space but it is not affect much of the discussion here. We have defined so called semi-curves in [10].

5.1.2 Mathematical Definitions of Discrete Deformation

Suppose there are two digital curves, A and B. Changing A to B "continuously" is equivalent to finding a sequence of digital curves in between A and B. Any two adjacent points in the two curves can be changed in minor ways. In computer graphics, such a process is called morphing. In mathematics it is called deformation [2].

Deformation means to change the shape of an object into another continuously by pressure or stress. For example, in Fig. 5.1a, curve A can be changed to curve B through a sequence of curves that bridge the two.

A

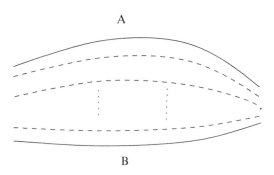

B

Fig. 5.1 An example of deforming curve A to B

For simplicity, we can explain the deforming process in the following way: we can consider a curve A as a function $A(x)$, $x \in R$. We can also consider a function $h_t(x)$ that associates A and B, where t represents time and $x \in R$. When $t = 0$, A is unchanged, so $h_0(x) = A(x)$. After some time, $h_1(x)$ becomes B.

Also we require $\lim_{\Delta \to 0}(h_t(x) - h_{t+\Delta}(x)) = 0$. This means that h_t is continuous at t. More formally, functions f, g are said to be deformed if there is a function H,

$$H : R \times [0,1] \to R \tag{5.1}$$

$H(x,0) = f(x)$ and $H(x,1) = g(x)$, where H(x,t) is continuous. This definition is somewhat more abstract for people familiar with mathematics. In general we may need to define a function H from one space (manifold) X to space (manifold) Y: $H : X \times [0,1] \to Y$.

Intuitively, a curve can be deformed to another if and only if there is a sequence of lofted curves between them.

However, in Fig. 5.2, curve C could not be changed to curve D since there is a hole between C and D. Whether a curve can be deformed to another depends on whether one can draw a sequence of curves in between without meeting any holes in the space.

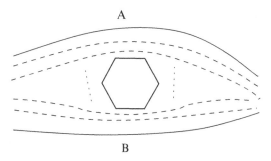

Fig. 5.2 Curve A can not be deformed to B since there is a hole between them

For digital and discrete cases, deformation is harder to define. We first review the continuity of the discrete space.

In Chap. 3, we defined digital continuous functions and gradually varied functions [6–8]. A digital "continuous" function is an integer function in which the value at a digital point is the same or almost the same as the value of its neighbors. In other words, if x and y are two adjacent points in a digital space, then $|f(x) - f(y)| \le 1$.

For the digital space and discrete space, we simply use digital and discrete manifolds to indicate the digitization or discretization of connected manifolds or objects. More formal definitions can be found in [10].

Digital (or discrete) deformation was first defined by Khalimsky in 1987 [17] and then modified by Boxer in 1999 [5]. Boxer's earlier work in [4] was based on Rosenfeld's paper [24].

Definition 5.1. Let X and Y be digital (or discrete) manifolds and $f,g : X \to Y$ be two digital continuous functions. g is said to be digitally deformed from f in Y if there is a consecutive integer set $N_m = \{0, 1, \cdots, m\}$ and a function $H : X \times N_m \to Y$ such that

(1) For all $x \in X$, $H(x,0) = f(x)$ and $H(x,m) = g(x)$;
(2) For all $x \in X$, the induced function $H_x : N_m \to Y$, defined by $H_x = H(x,t)$ and
    $t \in N_m$, is digitally continuous; and
(3) For all $t \in N_m$, the induced function $H_t : X \to Y$, defined by $H_t = H(x,t)$ and
    $x \in X$, is digitally continuous.

This definition is mainly a simulation of the definition in continuous space. It is very intuitive but has some problems in terms of completeness. This is because in continuous cases, there is no jump from one intermediate curve to another. However, in digital cases, there are only a few curves in between. How we eliminate holes between two consecutive curves is a problem. This problem can be solved easily using the geometric definition of deformations in the next section. However, more research is needed in terms of what should be done with an algebraic definition of digital deformation.

For 2D spaces (2D images), Rosenfeld defined local deformation for two digital curves C and D [21]: every pixel (point) of C coincides with or is a neighbor of a pixel (point) of D, and vice versa. This is an intuitive definition; however, it does not work in every case. It is close to a necessary condition.

Rosenfeld and Nakamura modified the definition by adding some restrictions so that the inside of C and outside of D are disjoint in order to avoid such a counter example. They called this strong local digital deformation [22]. This definition is sufficient but will miss some cases. In addition, this method is hard to deal with in 3D.

Herman (1998) simply used single surface-cell reduction/addition to help the definition of deformation, which is closer to the original meaning of topology. However, it is also restricted to 2D images. Herman's concept is called elementarily 1-equivalent. His definition is not considered well defined in terms of mathematics [16].

In fact, Newman in 1954 had already discussed similar cases in his seminal book for continuous space [19]. The central concern is to see whether a surface-unit (two-cell) surrounded by the union of two curves is inside the target space. The problem is that in continuous space, we define complexes. Even though all points on the boundary are in the target space, the inside face may not be in the target space. This is because if the complex is not a single point, it can still contain another point inside.

In digital cases, we only have finite points so if we define complexes, it will generate some ambiguity. We have to force the case to be included, meaning that if all corner points of a 2-cell is in the target space, then the 2-cell is in the target. That is to say, in order to define a digital curve, one must first define a digital surface [9, 10]. The deformation can be defined on moving a curve along with a different side of the 2-cells on the digital surface.

In Chen's modifications, two simple digital curves are considered gradually varied if their union contains no closed curves. This union can only contain semi-closed curves that are union of some 2-cells [10].

Based on the discussion above, Definition 5.1 is still the best in terms of being in a pure algebraic form. $N_m = \{1, 2, \cdots, m\}$ may not be connected without being defined as such, so a chain C has a better geometric meaning in terms of continuity than the integer set N_m. We make a little modification of Definition 5.1 as follows:

Definition 5.2. We say that the function H is a digital homotopy between f and g.

Let X and Y be digital or discrete manifolds. Two digitally continuous functions $f, g : X \to Y$ are said to be digitally homotopic in Y if there is a chain C with $(m+1)$ vertices named $0, 1, \cdots, m$ consecutively and a function $H : X \times C \to Y$ such that:

(1) For all $x \in X$, $H(x, 0) = f(x)$ and $H(x, m) = g(x)$;
(2) For all $x \in X$, the induced function $H_x : C \to Y$, defined by $H_x = H(x, t)$ and $t \in C$, is continuous; and
(3) For all $t \in C$, the induced function $H_t : X \to Y$, defined by $H_t = H(x, t)$ and $x \in X$, is continuous.

If a function follows (1)–(3), then we say that the function H is a digital homotopy between f and g.

In this definition, we are still unable to deal with the case where a cell is not in the manifold but its four corner points are in the manifold. In the next section, we simulate Newman's idea in discrete space to deal with this problem.

5.2 Geometric Deformation of Digital Curves

In differential geometry, any simple differentiable closed curve with its internal side in 2D has a conformal mapping to a disk [13]. But for digital cases, the resolution is usually fixed. It is hard to determine its conformal mapping without refining the grid.

On the other hand, it is true that we need an algebraic definition for deformation in digital or discrete space. It seems that it is not very easy to reach this goal.

However, we could get a relatively precise geometric definition for digital deformation. It may be a good idea for now in terms of implementations.

"Gradual variation" is still a good term that gives an intuitive meaning of the "continuous" change between two discrete simple paths. Herman first defined the concept of "elementarily N-equivalent" in 2D to describe changing a few two-cells in simply connected space [16]. We have given a variation of "elementarily 1-equivalent" in [10].

In [10], we call a simple path a semi-curve (pseudo-curve) since it could contain a two-cell. Intuitively, "continuous" change from one simple path C to another C' is defined as a change where there is no "jump" between these two paths. $x, y \in S$, then $d(x, y)$ denotes the distance between x and y. $d(x, y) = 1$ means that x and y are adjacent in S. The following definition was based on [10].

Definition 5.3. Two simple paths $C = p_0, \ldots, p_n$ and $C' = q_0, \ldots, q_m$ are gradually varied in 2D digital manifold S if $d(p_0, q_0) \leq 1$ and $d(p_n, q_m) \leq 1$. In addition, for any point p in C that is not an endpoint, we have

1. p is in C', or p is contained by a two-cell A (in S) such that A has a point in C', and
2. Each non-end-edge in C is contained by a two-cell B (in S), which has an edge contained by C' but not C if C' is not a single point.

These also hold for a non-end point q in C'.

In the above definition, having two-cells A and B belong in S is very important to the correctness of the definition. The following example explains this concern [10].

C and C' in Fig. 5.3a are gradually varied, but C and C' in Fig. 5.4b are not gradually varied.

We can see that a surface-cell, which is a simple path, and any point in the surface-cell are gradually varied. Assume $E(C)$ denotes all edges in path C. Let $XorSum(C,C') = (E(C) - E(C')) \cup (E(C') - E(C))$. $XorSum$ is called *sum* (*modulo2*) in [19].

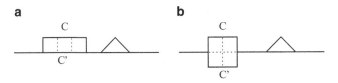

Fig. 5.3 (a) C and C' are gradually varied; (b) C and C' are not gradually varied

Lemma 5.1. *Let C be a pseudo-curve (simple-path) and A be a surface-cell. If $A \cap C$ is an arc containing at least one edge, then $XorSum(C,A)$ is a gradual variation of C.*

It is not difficult to see that

$$XorSum(XorSum(C,A),A) = C$$

and

$$XorSum(XorSum(C,A),C) = A$$

under Lemma 5.1.

Definition 5.4. Two simple paths (or pseudo-curves) C, C' are said to be homotopic if there is a series of simple paths $C_0, \ldots, C_n$ such that $C = C_0$, $C' = C_n$, and C_i, C_{i+1} are gradually varied.

Lemma 5.2. *If two simple paths C, C' are homotopic, then there is a series of simple paths $C_0, \ldots, C_m$ such that $C = C_0$, $C' = C_n$, and $XorSum(C_i, C_{i+1})$ is a surface-cell excepting end-edges of C, C'.*

In the rest of the section, we compare the geometric definition of deformation with the algebraic definition in the previous section. In other words, we put the geometric definition, Definitions 5.3 and 5.4, within the context of the definition, Definition 5.1, by Khalimsky and Boxer. We will see that the geometric definition satisfies all conditions of the algebraic definition.

By Definitions 5.3 and 5.4, for the beginning curve and the ending curve, the following is true in Definition 5.1: (1) For all $x \in X$, $H(x,0) = f(x)$ and $H(x,m) = g(x)$.

Since every simple path is digitally continuous in Definition 5.3 and 5.4, we also have that the condition (2) in Definition 5.1 is true: (2) For all $x \in X$, the induced function $H_x : C \rightarrow Y$, defined by $H_x = H(x,t)$ and $t \in C$, is continuous;

For the third condition in Definition 5.1, (3) for all $t \in C$, the induced function $H_t : X \rightarrow Y$, defined by $H_t = H(x,t)$ and $x \in X$, is continuous. There must be a trice at t (time) (from 0 to m) The length of C can be varied and one can insert some pseudo points to H_t, which is a simple path with respect to the variation of t.

However, to say that Definition 5.1 satisfies the geometric definition is difficult. We may need to use global "connection" as the media or put it into continuous space for the analogue case.

It is possible to find an example that satisfies all the conditions of Definition 5.1 but not a "continuous" deformation. However, Definition 5.1 is still very reasonable in terms of being an algebraic definition.

The problem of geometric definitions is that the simple path is not a "pure" curve in digital space, it may contain a two-cell. It is called a semi-curve in [10]. We call this state an unstable state. It must be self-reorganized to a stable state, which does not contain any two-cells. The smallest changes it needs to become a stable state is called the primary stable state. However this is not unique and other states can also be defined.

5.3 New Consideration of Discrete Deformation and Homotopy

The purpose of digital deformation is to apply the method to real world applications and to design algorithms. Deformation means a "continuous" change of the shape in digital space. The minimum distance in digital space is "1;" therefore, the digital deformation can be defined as a series of digital objects, $D_1, D_2, \ldots, D_n$, where any adjacent D_i and D_{i+1} have a "distance" equal to 1. This means that there is a mapping f from D_i to D_{i+1} such that $d(x, f(x)) \leq 1$ for all $x \in D_i$.

This intuitive idea is from Rosenfeld. Its problem is that D_i and D_{i+1} do not contain the same number of points so there will be some cases that will not meet this definition. In other words, this is the necessary condition of digital deformation, but it is not sufficient.

For a 2D digital space M, Herman suggests that changes be made one at a time. Chen modified this idea by combining it with Newman's method: $XorSum$ of D_i and D_{i+1} must be the union of two-cells in 2D digital space M.

We can extend such idea to any discrete space: Let D_i be a k dimensional object. There will be a partition P_j of $D_i \cup D_{i+1}$ such that in some, D_i in P_j is the same as D_{i+1} in Pj. In others, D_i in P_j and D_{i+1} in P_j form a $(k+1)$-cell.

This definition will generate another problem during the deformation. D_i may contain higher dimensional cells. This case differs from continuous space, and we call this an unstable state. Once the deformation is complete, the higher dimensional cells will collapse. Most definitions have overlooked this fact. The geometric definition may be formatted using functions by adopting algebraic technique. It is a question left to readers to think about.

For general k-manifolds, let X and Y be k-D digital (or discrete) manifolds in a K-D space where $(K > k)$. Two digitally continuous functions $f, g : X \to Y$ are said to be digitally homotopic in Y if there is a chain C with $(m+1)$ vertices named $0, 2, \ldots, m$ consecutively and a function $H : X \times C \to Y$ such that

(1) For all $x \in X$, $H(x, 0) = f(x)$ and $H(x, m) = g(x)$; (2) For all $x \in X$, the induced function $H_x : C \to Y$, defined by $H_x = H(x, t) and t \in C$, is continuous; and (3) For all $t \in C$, the induced function $H_t : X \to Y$, defined by $H_t = H(x, t) and x \in X$, is continuous.

The question now becomes whether we can eliminate some of these conditions: We want to prove that this definition is the same as the geometric definition of deformation.

5.4 Basic Concepts of Fundamental Groups and Algebraic Computing*

In this section, we overview the algebraic aspects of deformation related to topology and its applications. This section contains some profound knowledge in topology [2, 14]. We only give a brief overview.

5.4.1 Fundamental Groups

Fundamental groups, introduced by H. Poincare, is an essential concept connecting algebra and geometry. Two topological spaces are equivalent under the deformation and the cut-and-paste procedures. This is called homemorphism, which means there is a continuous invertible mapping between two spaces.

Two spaces that are topologically equivalent (homemorphic) implies they are homotopic. Thus, homotopic properties are topological invariants. Determining two spaces for homotopy is sometimes very difficult. It may not even be computable in general cases. Mathematicians found a way to determine whether two topologic spaces are not homotopic by finding some homotopic properties that do not hold in two different spaces. In other words, if two spaces are homotopic, they must have certain properties. If one of the properties does not hold, then these two spaces are not homotopic. Topologists usually call these properties invariants or characteristic invariants.

The fundamental group is the simplest invariant, but it is usually very difficult to obtain.

A *group* $G = (S, \cdot)$ is a simplest algebraic structure: a base set S and an operation $\cdot$ can be viewed as a binary function to S, $f : S \times S \to S$. The definition of a group G is as follows:

(a) If $a, b \in G$, then $a \cdot b \in G$ (Closure Property). (b) If $a, b, c \in G$, then $(a \cdot b) \cdot c = a \cdot (b \cdot c)$. (Associativity). (c) There exists an $e \in G$, such that for every $a \in G$, $e \cdot a = a \cdot e = a$ (Identity). (d) If $a \in G$, there exists $b \in G$ such that $a \cdot b = b \cdot a = e$ (Inverse Element).

If $a \cdot b = b \cdot a$, then this group is called an Abelian group (Commutative). In general, fundamental groups are not Abelian.

Let M be a topological space and $x \in M$ is a point; a closed curve starting and ending at x is called a loop $loop(x)$. Then, x would also be called a base point. Given a direction of the loop, clockwise or counterclockwise, the loop now has a path.

The loops with base point x will form a group when we define the product operations. The group product $a \cdot b$ of loop a and loop b is defined as the path of a followed by the path of b.

Two loops a and b with base point x are equivalent if a can be deformed to b. An identity element is the point x (a special loop). A loop a's inverse element is defined as the inversed path of a.

We can prove that all loops with base point x form a group $\Pi_x(M, \cdot)$, which is called the fundamental group. The fundamental group was defined by Poincare in 1895 [2]. For a (path-) connected space M where $x, y \in M$, $\Pi_x(M, \cdot)$ and $\Pi_y(M, \cdot)$ are equivalent (isomorphic) [2]. fundamental groups for digital spaces were first studied by Khalimsky [17], and then Kong [18]. The path or curves are formed by digital points. The definition is used to simulate the classic fundamental groups in digital space by defining the digital curves and their motions.

5.4.2 Homotopy Groups

A loop passing x in M can be viewed as a continuous function $f : [0, 1] \to M$ where $f(0) = f(1) = x$. We can also let X be a unit circle and have $f : X \to Y$ be a loop in Y.

The homotopy group is a general form of the fundamental group for higher dimensions. We can define it as an n-(unit)ball or n-(unit)sphere. For instance, $[0, 1]$ is a one-ball.

When X is an n-ball, in the above definition, there is a fixed point p in Y where H must pass through p. All homotopic mappings form a class, which is an n-homotopy group. If $n = 1$, the homotopy group is just the fundamental group. If $n > 1$, the group will be an Abelian group [14].

For an n-ball, the mapping will map the boundary of the n-ball to a point if we have an n-closed manifold in Y.

In discrete cases, a general definition of a simply connected space be: a digital space is simply connected if any two closed simple paths with a base point x are homotopic. Therefore, the fundamental group of a simply connected space has only one element.

5.4.3 Homology Groups and Genus Computing

Homology groups are used to detect and categorize holes in a manifold. Its definition is even more difficult than homotopy groups and it needs knowledge of abstract algebra [14]. Fortunately, for a connected (and orientable) surface, the genus g also indicates the number of handles or tunnels on the surface. This value has fundamental importance in topology.

In general, the polyhedral surfaces are called two-dimensional finite CW- complexes, which is made of basic building blocks usually called two-cells. For example, combining points, one-cells, and will form a simplicial complex. However, cubic space is made by squares or rectangles.

Euler characteristics is topological invariant that is defined for any finite CW-complex:

$$\chi = k_0 - k_1 + k_2 - k_3 + \cdots,$$

where k_i indicates the number of cells in dimension i in the complex. We know that

$$\chi = 2 - 2g - b$$

for surfaces with b boundary components. So $\chi = 2 - 2g$ for closed surfaces.

In digital space, since the Euler characteristic can be calculated based on the number of i-cells, we can obtain genus g by using $\chi = 2 - 2g - b$. The calculation complexity of χ needs to consider all k-cells in the manifold, $k = 0, \ldots, m$.

An even simpler formula related to homology groups in 3D digital space was obtained by Chen and Rong in [11]. This theorem states: Let M be a closed digital 2D manifold in direct adjacency (i.e. a (6,26)-surface in 3D). The formula for genus is

$$g = 1 + (M_5 + 2M_6 - M_3)/8. \tag{5.2}$$

where M_i indicates the set of surface-points, each of which has i adjacent points on the surface. This formula is important to understanding the structure of 3D images and has been used in 3D graphics [12] If M is simply connected, i.e. $g = 0$, then $M3 = 8 + M_5 + 2M_6$ [10].

5.5 Case Study: Rosenfeld's Problem of Digital Contraction

Contraction is a specific deformation procedure that usually deforms a curve to a simple point. It also means reducing the size of the original object to make it smaller. In 1996, Rosenfeld initiated a procedure to digital contraction . It seemed like Rosenfeld had not completed the development of the algorithm for digital contraction on a 2D closed digital manifold. However, he had presented a very valuable and remarkable idea. We present his idea here to restate this open problem.

The center of Rosenfeld's technique is the local contraction. The purpose of this subsection is to give an example of the digital contraction. The main resource was a paper by Rosenfeld in 1996 [21]. We redo the surface-cell reduction to realize the Rosenfeld's original method.

In Rosenfeld's definition, C is said to be digitally contractible to a point in $S \subset \Sigma_2$ if there exists a sequence of curves $C_0, C_1, \cdots C_n$ all in S, $C = C_0$, and C_n is a simple point (a trivial closed curve), such that C_{i+1} is a local contraction of C_i, $i = 1, \cdots, n$.

Using the definition of surface-cells in Sect. 5.3, we can rewrite the digital contraction as: Let C be a simple closed path in Σ_2. C is also called a (pseudo-) digital curve. Let $A(C)$ be the point set of the inside region bounded by C, $C \subset A(C)$. A digital curve D is said to be a local contraction of C if $D \subset A(C)$ and D is a gradually varied deformation of C (the deformation does not contain a jump). Figure 5.4 shows three digital curves.

```
a                b                  c
-  -  -  -  -    -  -  1  1  1      -  -  1      -
-  1  1  1  -    -  1  1  -  1      -  1      1  -
-  1  -  1  -    -  1  -  1  1      -  1      1  -
-  1  1  1  -    -  1  1  1  -      -  -  1  -  -
-  -  -  -  -    -  -  -  -  -      -  -  -  -  -
```

Fig. 5.4 Closed digital curves: (**a**) and (**b**) four-connected curves, (**c**) a eight-connected curve

```
a                              b              c
-  -  -  -  -   -  -  -  -  -   -  -  -  -     -  -  -  -
-  c  1  1  -  -    c  c  -  -  -  -  -  -     -  -  -  -
-  1  d  1  -  -  1  d  1  -  -  1  d  c  -   -  1  1  -  -   -  -  1  -  -
-  1  1  1  -  -  1  1  1  -  -  1  1  c  -   -  1  1  -  -   -  -  -  -
-  -  -  -  -  -  -  -  -  -  -  -  -  -  -   -  -  -  -     -  -  -  -
```

Fig. 5.5 Digital curve contraction: (**a**) and (**b**) contraction steps, (**c**) the final result

In Fig. 5.5, "1" can be included in both C and D, where c is an element of C and d is an element of D. Therefore, the purpose of the above process is to reduce the number of two-cells in $A(C)$. We can see that the reduction of two-cells from Fig. 5.5(a) to Fig. 5.5(b). Then we get Fig. 5.5(c), a single point.

5.6 The Difference Between Discrete Deformation and Continuous Deformation

Continuous deformation, we usually assume that the manifold does not contain any infinitely small holes or tunnels.

In other words, the diameters of the holes or tunnels must be constant. Otherwise, the continuously moving function will find a jump even if the hole is very small.

However, in digital cases, each move is at a distance of "1." If there is a hole or tunnel with a distance of less than 1 (for example a two-cell without the boundary), then the digital move will not be able to find the jump. Most definitions overlook this fact, which is why we have to say that the unit of a digital space must be smaller than a hole. In other words, a closed curve must contain eight points and not four.

```
a                     b
-  -  -  -  -        -  -  -  -  -
-  1  1  1  -        -  1  1  1  -
-  1  -  1  -        -  1  H  1  -
-  1  1  1  -        -  1  1  1  -
-  -  -  -  -        -  -  -  -  -
```

Fig. 5.6 Contractible or uncontractible digital curves: (**a**) A contractible curve (**b**) H is a hole not in the set S; The digital curve is not contractible

We said that Definition 5.1 is not perfect since such a problem must be resolved in the algebraic definition, even though we do not know how it can be done. This research contains many unclear and unsolved problems. However, it may contain great potential in solving some significant homotopy group problems in the future.

In the discretization method, quantization (digitization) is used first before marching cubes to obtain the initial triangulation. After that, triangles can be combined to make larger triangles.

What is the drawback of continuous deformation? In general, continuous deformation only applies to specific spaces where the space can be defined using formulas or equations. Otherwise, how can anyone define a sequence of continuous mappings on the general manifold in terms of algorithms or procedures?

5.7 Remarks

For homotopy groups, the calculation seems to require the use of continuous functions in discrete space. Digitally continuous functions can play a major role in digital images. Also see [19].

For more theoretical results about topology in algebra, refer to Hatcher's book [14]. For 3D image boundary tracking, refer to [3]. For a special contraction related to gradually varied mapping in graph theory, refer to [1].

References

1. Agnarsson G, Chen L (2006) On the extension of vertex maps to graph homomorphisms. Discrete Math 306(17):2021–2030
2. Alexandrov PS (1998) Combinatorial topology. Dover, New York
3. Brimkov V, Klette R (2008) Border and surface tracing. IEEE Trans Pattern Anal Mach Intell 30(4):577–590
4. Boxer L (1994) Digitally continuous functions. Pattern Recognit Lett 15(8):833–839
5. Boxer L (1999) A classical construction for the digital fundamental group. J Math Imaging Vis 10(1):51–62
6. Chen L (1990) The necessary and sufficient condition and the efficient algorithms for gradually varied fill. Chinese Sci Bull 35:10
7. Chen (1991) Gradually varied surfaces on digital manifold. In: Abstract of international conference on industrial and applied mathematics, Washington, DC, 1991
8. Chen L (1994) Gradually varied surface and its optimal uniform approximation. In: IS&TSPIE symposium on electronic imaging, SPIE Proceedings, vol 2182 (Chen L, Gradually varied surfaces and gradually varied functions, in Chinese, 1990; in English 2005 CITR-TR 156, U of Auckland. Has cited by IEEE Trans in PAMI and other publications)
9. Chen (L) (1999) Note on the discrete Jordan curve theorem. In: Vision geometry VIII, Proceedings SPIE, vol 3811, Denver
10. Chen L (2004) Discrete Surfaces and Manifolds: a theory of digital-discrete geometry and topology. SP Computing
11. Chen L, Rong Y (2010) Digital topological method for computing genus and the Betti numbers. Topol Appl 157(12):1931–1936
12. Etiene T, Nonato LG, Scheidegger C, Tierny J, Peters TJ, Pascucci V, Kirby RM, Silva C (2012) Topology verification for isosurface extraction. IEEE Trans Vis Comput Graph 6(18):952–965
13. X. Gu and S-T Yau, Computational Conformal Geometry, International Press, Boston, 2008.
14. Hatcher A (2002) Algebraic topology. Cambridge University Press, Cambridge/New York
15. Han SE (2005) Digital coverings and their applications. J Appl Math Comput 18(1–2): 487–495
16. Herman GT (1993) Oriented surfaces in digital spaces. CVGIP 55:381–396.
17. Khalimsky E (1987) Motion, deformation, and homotopy in finite spaces. In: Proceedings IEEE international conference on systems, man, and cybernetics, pp 227–234. Chicago
18. Kong TY (1989) A digital fundamental group. Comput Graph 13:159–166
19. Newman M (1954) Elements of the topology of plane sets of points. Cambridge University Press, London
20. Rosenfeld A (1986) Continuous' functions on digital pictures. Pattern Recognit Lett 4:177–184
21. Rosenfeld A (1996) Contraction of digital curves, University of Maryland's Technical Report in Progress. ftp://ftp.cfar.umd.edu/TRs/trs-in-progress/new.../digital-curves.ps
22. Rosenfeld A, Nakamura A (1997) Local deformations of digital curves. Pattern Recognition Letters, 18:613–620

Part II
Digital-Discrete Data Reconstruction

Chapter 6
Basic Numerical and Computational Methods

Abstract The most popular problem in data reconstruction is fitting a curve or a surface. Data fitting has two meanings: interpolation and approximation. Interpolation means fitting the data exactly on the sample values while approximation means a fitted function can be set near the sample values. In this chapter, to introduce data reconstruction, we do a comprehensive review on basic numerical and computational methods for data fitting. We will cover the the following topics: (1) Piece-wise linear interpolation and approximation; (2) Delaunay triangulation; (3) Lagrange curve interpolations and smooth curve approximation; and (4) Coons surfaces and spline surface fitting. In curve and surface fitting, we briefly introduce some of the advanced methods including B-Spline and Bernstein polynomial approximation.

6.1 Linear and Piecewise Linear Fitting Methods

Given a set of sample points along with a line (e.g. x-axis), if we want to fit a curve on these points, the simplest and probably most effective way is to link these points together. The word "link" describes connecting two sample points with a straight line. This method is called a piecewise linear interpolation.

Mathematically, a straight line, or linear interpolation, between two points (x_0, y_0) and (x_1, y_1) can be derived by a slope equation: Adding an x between x_0 and x_1, $p = (x, y)$ shall follow the slope equation as follows:

$$\frac{y - y_0}{x - x_0} = \frac{y_1 - y_0}{x_1 - x_0}. \tag{6.1}$$

So,

$$y = y_0 + (x - x_0)\frac{y_1 - y_0}{x_1 - x_0}. \tag{6.2}$$

For two dimensional cases, given three points in space, one can draw three line segments, a triangle, on the three sample points. This method is very popular in

L.M. Chen, *Digital Functions and Data Reconstruction: Digital-Discrete Methods*,   67
DOI 10.1007/978-1-4614-5638-4_6, © Springer Science+Business Media, LLC 2013

computer graphic applications. Therefore, the simple line equation (6.2) provides a basis to piecewise-linear curve and surface interpolations.

Not every data-fitting problem can be solved using such a simple interpolation, for example, in the case of a set of random sampled data points including noise or errors. We must apply an approximation method of finding a function to best fit this data set. Linear regression is one such method that obtains a line (or surface) using the minimum total distance to sample points.

The main idea of linear regression is to assume a linear equation, an un-determined line,

$$y = a + b \cdot x \tag{6.3}$$

and then to get the values of a and b based on the sample points. The technique of solving this problem is called the least squares method that has a very general importance in data fitting. We discuss it further in the case study section of this chapter.

6.1.1 Piecewise Linear Method

Not every desired function is a line; sometimes we need a piecewise linear function to approach a curve, the easiest way is to use line-segments to approximate it. This function is usually for obtaining continuous functions that do not need to be differentiable. It is one of the most important and practical fitting methods. See Fig. 6.1.

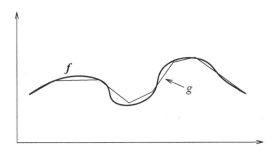

Fig. 6.1 Curve f and its approximation by piecewise linear function g

For a sequence of n sample points (x_i, y_i) with $x_1 < \cdots < x_n$, the equation consists of a set of linear functions as given in (6.2):

$$y = y_i + (x - x_i)\frac{y_{i+1} - y_i}{x_{i+1} - x_i} \ , x \in [x_i, x_{i+1}] \tag{6.4}$$

This is valid since the slopes of a line segment separated by unknown point (x, y) are the same.

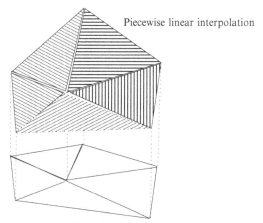

Domain triangulation by 6 sample points

Fig. 6.2 Domain decomposition and its linear interpolation by triangles

For 2D cases, a piecewise linear surface can also be made simple. The core of the reconstruction is to first use domain decomposition: project the sample points to the *xy*-plane, then get a triangulation of those sample points on the plane so we have a set of triangles whose vertices are the projected sample points. Next, we restore the height (value on the *z*-axis) of all the sample points. So the new triangles will be the piecewise linear surfaces in the problem. Fig. 6.2 gives an example.

To decompose a 2D domain, the Voronoi diagram and Delaunay triangulation are most commonly used methods. This is because the Voronoi diagram provides a natural way of partitioning a domain: a point is categorized to belong to a sample point (a site) if it is closest (it has the shortest distance) to the point compared to other sample points.

A Voronoi diagram is made by some Voronoi regions; each Voronoi region is a polygon. See Fig. 2.4. The Delaunay triangulation is shown using dashed lines made by lining two sample points if their Voronoi regions share a polygon edge. We will give an algorithm in the case study for obtaining the Delaunay triangulation.

6.1.2 Case Study 1: Least Squares for Linear Regression

Assume we have n sample data points (x_i, y_i). These points are on a straight line, but the sampling process might have some observation errors. Based on the equation of a line (6.2), we have:

$$y_i = a + bx_i, \ i = 1, \cdots, n \tag{6.5}$$

However, since there are observation or sampling errors, this equation may not always hold true for all sample points. In other words, $e_i = y_i - (a + bx_i)$ is not always zero. What we can do is find a and b such that the summation of errors is minimized,

i.e.
$$\text{Minimize } E = \Sigma_{i=1}^{n} |e_i|$$

where E is the function of a and b so that $E = E(a,b)$. To get the minimum $E(a,b)$, according to the extreme value theorem in calculus, the following two equations must hold:

$$\frac{\partial E}{\partial a} = 0 \ \& \frac{\partial E}{\partial b} = 0$$

Since the derivative of an absolute value function is hard to calculate, and it does not have a continuous derivative, we use the square of the error, $(e_i)^2$, to replace $|e_i|$. Therefore,

$$\text{Minimize } E = \Sigma_{i=1}^{n} (e_i)^2 \tag{6.6}$$

This equation minimizes the square of errors, the least square method .

$$\frac{\partial E}{\partial a} = \frac{\partial}{\partial a}(\Sigma(y_i - (a+bx_i))^2) = \frac{\partial}{\partial a}(\Sigma - 2(y_i - (a+bx_i)))$$

$$\frac{\partial E}{\partial b} = \frac{\partial}{\partial b}(\Sigma(y_i - (a+bx_i))^2) = \frac{\partial}{\partial b}(\Sigma - 2x_i(y_i - (a+bx_i)))$$

Thus, we have a system of linear equations:

$$na + (\Sigma x_i) \cdot b = \Sigma y_i$$

$$(\Sigma x_i) \cdot a + (\Sigma x_i^2) \cdot b = \Sigma(x_i \cdot y_i)$$

We can easily solve the equations using computers. It is slightly complex to represent in the following explicit format:

$$a = \frac{(\Sigma_{i=1}^{n} y_i \Sigma_{i=1}^{n} x_i^2) - (\Sigma_{i=1}^{n} x_i \Sigma_{i=1}^{n} x_i y_i))}{n\Sigma_{i=1}^{n} x_i^2 - (\Sigma_{i=1}^{n} x_i)^2} \tag{6.7}$$

$$b = \frac{(n\Sigma_{i=1}^{n} x_i y_i) - \Sigma_{i=1}^{n} x_i \cdot \Sigma_{i=1}^{n} y_i)}{n\Sigma_{i=1}^{n} x_i^2 - (\Sigma_{i=1}^{n} x_i)^2} \tag{6.8}$$

Since all applications of the least squares method are under the same principle, this practice is essential to numerical analysis [3].

Some applications need to combine both the least squares and piece-wise linear methods. This is called segmented linear regression and may require the use of linear regression with artificial intelligence. This method will partition the domain into segments and then apply linear regression in each segment (See Fig. 6.3). People can also use an iterated method to dynamically segment the domain. However, this technique is still relatively new to researchers (www.waterlog.info/segreg.htm).

6.1.3 Case Study 2: Algorithm for Delaunay Triangulation*

There are many efficient algorithms [2] for the Voronoi diagram and Delaunay triangulation that were introduced in Chap. 2. In this subsection, we present the simplest one called the Bowyer-Watson algorithm. It is also very popular in practice [4].

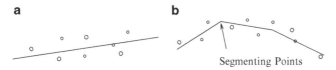

Fig. 6.3 Linear regression: (**a**) Linear regression, (**b**) Segmented linear regression

Another definition of Delaunay triangulation, for a set of sample points in 2D, is a triangulation such that no sample point is located inside the circumcircle of any triangle (in the triangulation). Therefore, the key idea of the algorithm is to use the circumcircle of a triangle to determine if the old triangle needs to be reformed.

Algorithm 6.1 (The Bowyer-Watson Algorithm). The Bowyer-Watson algorithm is a method for computing the Delaunay triangulation of a finite set of points in any number of dimensions. This algorithm works by adding one point at a time to a Delaunay triangulation of a subset of the original sample points. When the new added point is in a circumcircle, then we reconstruct the triangulation for a new Delaunay triangulation. The Bowyer-Watson algorithm is incremental:

Step 1: Insert a point, any existing triangles whose circumcircles contain the new point are marked as invalid.
Step 2: The invalid triangles are removed. This process leaves a convex polygon hole that contains the new point.
Step 3: Link the new point to each corner point of the convex polygon to form a new triangulation.

To obtain a Voronoi diagram of the points, we need the dual graph of the Delaunay triangulation, meaning that a perpendicular line is drawn to find the middle point of each triangle edge. These lines will stop when they intersect with two or more other lines at a point. The line segments are edges of Voronoi regions.

6.2 Smooth Curve Fitting: Interpolation and Approximation

In the computational sciences, interpolation and approximation are methods of constructing new data points based on sample data points. Interpolation does not replace the original sample points. On the other hand, in the approximation method, we may replace the original sample points using a new calculated function. In Sect. 6.1, we showed the linear interpolation and the least square approximation.

For a 1D domain or a single variable function, the Lagrange polynomial is usually used for theoretical purposes: the Lagrange polynomial guarantees a smooth interpolation exists. However, the Lagrange polynomial is not practical since it generates an n-degree polynomial on $n+1$ sample points that make the function unstable (a small change in guiding points may result in a large difference in the fitted function). As a result, the most popular numerical methods are the Bezier polynomial and B-spline methods.

6.2.1 Lagrange Interpolations

Lagrange interpolations give a simple formula for the existence of differentiable functions for finite points on a 1D domain. Polynomial interpolation is a generalization of linear interpolation. Note that the linear interpolant is a linear function. We now replace this interpolant by a polynomial of higher degrees.

Lagrange interpolation provides a general method for the existence of a smooth function that always exists for any number of guiding points. The disadvantage is that this polynomial has a very high degree, equal to the number of samples n. Such a high degree of smoothness may be unnecessary.

For a set of $m+1$ data points $(x_0,y_0),\ldots,(x_i,y_i),\ldots,(x_m,y_m)$ where $x_0 < x_1 < \ldots, < x_m$, the Lagrange interpolation polynomial is given by:

$$P(x) = \sum_{i=0}^{m} y_i \cdot P_i(x) \tag{6.9}$$

where,

$$P_i(x) = \prod_{0 \le j \le m;\ i \ne j} \frac{x - x_j}{x_i - x_j} \tag{6.10}$$

is called the Lagrange basis polynomial.

We can examine $P(x_i) = y_i$; it shows that this method interpolates the function exactly. However, Lagrange interpolation is usually only good in theory. Lagrange interpolation does not fit real world problems in many cases since it is too smooth and any small error will result in a big change in fitted curves.

The following example shows how a difference in a sample point will generate a huge change globally.[1] In Fig. 6.4, the original 12 points are in a straight line and when the middle point changes its y-axis location, this yields a big change for the entire curve. This is related to the so called oscillatory behavior or Runge's phenomenon [3, 5].

---

[1] Thanks to Mr. Mark Hoefer and Dr. Kirby Baker for the software at http://www.math.ucla.edu/~baker/java/hoefer/Lagrange.htm

Fig. 6.4 Lagrange interpolation : Original plot of sample points results in a baseline. A little boost in the middle point will make a big change at both ends

In other words, Lagrange interpolation does not preserve local properties in which a local chance only affects the function locally. This can usually be done using a spline function.

6.2.2 Bezier Polynomials

Bezier polynomials are often used in computer aided geometric design such as the automobile industry. The characteristic of the Bezier curve is: (1) It is an approximation method, and (2) the fitted curve is in the convex of the sample points (called control points). Bezier polynomials use the control points to manage the shape of the curve. It usually goes through two end points and does not pass through other control points.

Given a set of $m+1$ data points $P_0 = (x_0, y_0), P_1 = (x_1, y_1), \ldots, P_m = (x_m, y_m)$, Bezier polynomials will be calculated by parametric curve, e.g. $P(t) = (x(t), y(t))$, which is in vector form. For two points, the Bezier polynomial is given by:

$$P(t) = (1-t)P_0 + tP_1 \ , t \in [0,1] \tag{6.11}$$

In general, let $P_{\{0\ldots k\}}$ denote the fitted function for control points $P_0, \ldots, P_k$.

$$P(t) = P_{\{0\ldots m\}}(t) = (1-t) \cdot P_{\{0\ldots m-1\}}(t) + t \cdot P_{\{1\ldots m\}}(t) \ , t \in [0,1] \tag{6.12}$$

We can see that this formula is very similar to the formula for the linear interpolation of two points, but the reference "points" are the two lower order Bezier polynomials. In other words, Bezier polynomial fitting is defined as linear approximation recursively.

For the explicit form, we can expand the above equation to be

$$P(t) = \sum_{i=0}^{n} \binom{n}{i} (1-t)^{n-i} t^i \cdot P_i, \tag{6.13}$$

where $\binom{n}{i} = \frac{n!}{i! \cdot (n-i)!}$ is a binomial coefficient. A Bezier polynomial is also called a Bernstein polynomial of degree n. It can also be represented using a so called Bernstein basis polynomial of degree n, which is defined as

$$\beta_{i,n}(t) = \binom{n}{i} t^i (1-t)^{n-i}, \quad i = 0, \ldots, n. \tag{6.14}$$

Therefore,

$$P(t) = \sum_{i=0}^{n} \beta_{i,n}(t) \cdot P_i \tag{6.15}$$

The Bernstein polynomials is an "implementation" of the Weierstrass approximation theorem in Chap. 2. In practical computing and realization of Bernstein polynomials, we can select n points as samples, then double the sample points in next round. More discussion will follow in Chap. 7.

6.2.3 Natural Splines and B-Spline

For data fitting, the Lagrange polynomial and Bezier polynomial are high-degree polynomials. Even though they are smooth functions, high-degree polynomials usually have the oscillatory behavior, meaning that small local variations may cause a large error in the whole function.

The way to overcome the problem of over smoothness that occurs with piecewise-linear fitting and high-degree fitting is to have a lower degree polynomial. This is the idea behind splines.

Splines are used to smoothly link some smooth pieces together. Each smooth piece only fits a few points. For instance, four point data can fit a degree three polynomial called a cubic spline [3, 7]. Then, these splines are linearly combined to make the function.

6.2.3.1 Natural Cubic Splines

Cubic splines are the most popular type of spline functions. We will use the nonparametric method in this subsection and a parametric method in the next subsection.

First, we select the general cubic function as follows:

$$f(x) = a + bx + cx^2 + dx^3 \tag{6.16}$$

To determine four coefficients a, b, c, d, we need four sample points. Then, we need a system of linear equations to obtain these four values. The reconstructed curve will pass through the four points. (It is obvious that this function is similar to Lagrange.) If we have more than four sample points, say five or six sample points, then we need two or more curves, each of which are made by four sample points. These curves will then be joined together.

More importantly, the joint locations need to be smooth (C^2). The joint location is used to separate two segments in domain. Each curve made by the four points (a function on the segment) is called a spline. The original meaning of a spline comes from deformable wooden sticks.

Let us assume that we have $n+1$ sample points. A natural cubic spline uses each segment $[x_i, x_{i+1}]$ in the x-axis to build a spline:

$$f(x) = a_i + b_i \cdot x + c_i \cdot x^2 + d_i \cdot x^3 \text{ for } x_{i-1} \le x \le x_i, i = 1, 2, \cdot, n.$$

So we have total n splines with $4 \cdot n$ unknown coefficients. To solve these $4 \cdot n$ unknown variables, we need $4 \cdot n$ linear equations (constraints). As we know, we require the following conditions to be true:

(1) Interpolation condition: $f(x_i) = y_i$; $i = 0, 1, \ldots, n$
(2) Continuous Condition: $f(x_i + \delta) = f(x_i - \delta)$; $i = 1, \ldots, n-1$
(3) First derivative continuous: $f'(x_i + \delta) = f'(x_i - \delta)$; $i = 1, \ldots, n-1$
(4) Second derivative continuous: $f''(x_i + \delta) = f''(x_i - \delta)$; $i = 1, \ldots, n-1$

Therefore, the total constraints from (1) to (4) are $(n+1) + 3(n-1) = 4n - 2$ and there are still two degrees of freedom for choosing the coefficients. The selection methods can be the following: (a) The natural cubic spline: $f''(x_0) = f''(x_n) = 0$, i.e. the second derivative of the first node and the last node are both 0, and (b). The "not-a-knot" condition: $f^{(3)}(x_1 + \delta)$ is continuous at x_1 and x_{n-1}. This cubic spline can be extended to any degree. However, in practice, it is the best as is [5, 6].

6.2.3.2 B-Spline*

Natural cubic splines are an interpolation technology. B-spline is an advanced method for curve approximation. In this section, we select a parametric method to introduce this method. In order to make the material easy to understand, we still maintain cubic B-splines as our focus.

In the context of natural cubic splines, we could use the Bezier cubic polynomial instead of a general polynomial. For the four sample points, P_0, P_1, P_2, and P_3, a Bezier cubic polynomial would pass through the first and last points.

The following is the most difficult issue to understand with the B-spline and also the best part of this technology. Since the middle two sample points, P_1 and P_2, are not on the fitted curve, so the segment on the curve is not very clear. This gives us the flexibility to assign pseudo-segments of sample points. Here, the sample points are called guiding points or control points since the final curve will most likely not pass through these points.

Given a set of $m+1$ data points $P_0 = (x_0, y_0)$, $P_1 = (x_1, y_1), \ldots, P_m = (x_m, y_m)$. Let $B(t) = (x(t), y(t))$, $t \in [0, 1]$ be the fitted curve.

Since four points will generate a cubic spline, we may have $m - (4)$ different cubic Bezier curves $P_0, \ldots, P_3; \ldots ; P_{m-3}, \ldots, P_m$.

How many pseudo-segments we can have? There are $m - 4$ inside the curve. However, if we need the curve to pass through the first and last points, we need to collapse points at the end points. So we will have $m + 4 - 1$ segments in $m + 4$ locations. These locations are called knots, with respect to t.

Therefore, each control point only affects four segments. This locality is a good feature of the approximation. Each Bezier polynomial is built on four knots (three segments of a t-axis).

The location of knots in the t-axis can vary. Then the curve will be different even though the control stays the same. That gives the auto industry, for example, a great deal of advantages in having cars with different body shapes.

After understanding the nature of B-splines, we can easily get the detailed formula for B-splines. It is the following:

Let the knot sequence be defined $\{t_0, t_1, \ldots, t_m\}$, where $t_0 \leq t_1 \leq, \ldots, \leq t_m$, $t_i \in [0, 1]$, and the control points are $P_0, \ldots, P_n$. Define the degree as $d = m - n - 1$. For instance, $d = 3$ is the cubic spline. (The knots $t_{(d+1)}, \ldots, t_{(m-d-1)}$ refer to internal knots. The fitted curve among these knots will maintain the properties desired.)

Define the basis functions as

$$B_{i,0}(t) := \begin{cases} 1 & \text{if } t_i \leq t < t_{i+1} \\ 0 & \text{otherwise} \end{cases}, i = 0, \ldots, m-2 \qquad (6.17)$$

$$B_{i,j}(t) := \frac{t - t_i}{t_{i+j} - t_i} B_{i,j-1}(t) + \frac{t_{i+j+1} - t}{t_{i+j+1} - t_{i+1}} B_{i+1,j-1}(t) \qquad (6.18)$$

where $i = 0, \ldots, m - j - 2$. $1 \leq j \leq d$ indicates the degree where i is the interval segment index or knots. B-spline is a linear combination of basis B-splines $B_{i,d}$ with coefficients P_i

$$\mathbf{P}(t) = \sum_{i=0}^{m-d-2} \mathbf{P}_i B_{i,d}(t), t \in [t_d, t_{m-d-1}].$$

We can see that this has a linear interpolation basis where the function is linearly accumulated p times.

Just like Bezier polynomial, this basis function iterates linearly d times to make d-degree smoothness. Therefore, the B-spline is a generalization of the Bezier curve. (If we let $m = d + 1$, $t_0 = \ldots = t_d = 0$, and $t_{d+1} = \ldots = t_{2d} = 1$, the Bspline becomes a Bezier curve.)

Some other nice properties can be used in practice: (1) adjusting the position of the knots will change the shape of the basis function, so we can adjust accordingly; (2) the fitted curve is contained in the convex hull of the control points.

In engineering and science, there is often a number of data points obtained by sampling or experimentation, which represent the values of a function for a limited number of values of the independent variable. It is often required to interpolate (i.e. estimate) the value of that function for an intermediate value of the independent variable. This may be achieved by curve fitting or regression analysis.

6.3 Numerical Surface Fitting

The classical method for surface reconstruction uses tensor product for smooth surface fitting. We first fit the four boundary curves in a rectangle area. Then, we use the boundary to fit the inside of the rectangle. Recently, researcher started to use moving least squares to fit the surfaces for random points (sometimes called cloud points). However, this technique must be based on the density of sample points.

We have discussed the triangulation method and the gradually varied method for continuous surface fitting in Chap. 3 and the beginning of this chapter. In the rest of this chapter, we introduce several methods for smooth surface fitting. Beginning in Chap. 7, we focus on the digital-discrete method for smooth surface fitting.

If we know that the collected data set is from a second-order equation in 3D space, we can use the quadratic surface equation as the basis function, before using the least squares method by finding the coefficients of the equation. A second-order surface in space has the general form:

$$x^2 + a_1 y^2 + a_2 z^2 + a_3 yz + a_4 zx + a_5 xy + a_6 x + a_7 y + a_8 z + a_9 = 0. \qquad (6.19)$$

This equation has nine coefficients; therefore, nine sample points will be enough to solve the equation. This is done by establishing nine linear equations before solving. If there are more than nine sample points, the least squares method, see Sect. 6.1.2, can be selected to get a best approximation. Another easy method is to randomly select nine points as a set to fit the equation and then find the average of the fitted surfaces of these selected sets.

The simple method above may not always be applicable to cases where we do not know the function of a surface. We now introduce Coons surfaces and B-splines for surface reconstruction.

6.3.1 Coons Surfaces

Coons surfaces are a natural way to fit a surface based on the four boundary curves: u_0, u_1, v_0, and v_1. Think about a middle curve between u_0 and u_1,

$$u(x, 1/2) = 1/2 u_0(x) + 1/2 u_1(x) \; ; x \in [0,1] . \qquad (6.20)$$

The lofted curve or morphing from u_0 to u_1 would be

$$u_y(x,y) = (1-y) \cdot u_0(x) + y \cdot u_1(x) \; ; x \in [0,1] . \qquad (6.21)$$

For v_0 and v_1, we would have

$$u_x(x,y) = (1-x) \cdot v_0(y) + x \cdot v_1(y) \; ; y \in [0,1] . \qquad (6.22)$$

Since the two equations above refer to the same function, we could use the average to represent the function.

$$u(x,y) = \frac{1}{2}[(1-y) \cdot u_0(x) + y \cdot u_1(x) + (1-x) \cdot v_0(y) + x \cdot v_1(y)] \; ; x, y \in [0,1] \quad (6.23)$$

Another way is to use the four corner points to estimate a reference surface, called the bilinear method, since we can use two corner points to get the boundary curve

$$u'_0(x) = (1-x)u_0(0) + xu_0(1) \text{ and } u'_1(x) = (1-x)u_1(0) + xu_1(1).$$

Therefore, a reference surface can be:

$$u_B(x,y) = (1-y)u'_0(x) + yu'_1(x)$$
$$= (1-y)(1-x)u_0(0) + (1-y)xu_0(1) + y(1-x)u_1(0) + yxu_1(1).$$

Now, we have three surfaces $u_x(x,y)$, $u_y(x,y)$, and $u_B(x,y)$ to choose from; the Coons surface selects a surface with bilinear blending (two similar surfaces are added and one surface made of corners is subtracted) [3],

$$u(x,y) = u_x(x,y) + u_y(x,y) - u_B(x,y)$$

Thus,

$$u(x,y) = (1-x) \cdot v_0(y) + x \cdot v_1(y) + (1-y) \cdot u_0(x) + y \cdot u_1(x) - u_B(x,y) ; \quad (6.24)$$

where $x,y \in [0,1]$.

The Coons surface applies to a rectangular area when four edges are known. This is the case in Fig. 1.1. When a region is decomposed into small rectangles, Coons surface will be an appropriate method to use. However, it does not guarantee that neighboring rectangles will be smooth. The next subsection will introduce a tensor product for splines for this purpose.

6.3.2 Bezier and Bspline Surfaces with Tensor Products*

Bezier and B-spline surfaces are constructed by two Bezier or Bspline curves to make a surface-area or patch, which is called the tensor product.

$$S(u,v) = \sum_{i=0,n} \sum_{j=0,m} B_{i,n}(u)B_{j,m}(v)P_{i,j} , u,v \in [0,1] . \quad (6.25)$$

where $P_{i,j}$ are $(n+1)(n+1)$ control points (sample points) and $B_{i,n}(u)$ or $B_{j,m}(v)$ can be either Bezier polynomials or B-spline basis functions discussed in the above sections.

Some advanced surface reconstruction techniques are also based on tensor products. For instance, non-uniform rational B-splines (NURBS) are very common today in computer graphics in controlling flexible shapes. Non-uniform means that a different weight must be added to each spline.

A non-uniform rational basis function is defined as

$$R_{i,j}(u,v) = \frac{B_{i,n}(u)B_{j,m}(v)w_{i,j}}{\sum_{s=0}^{n} \sum_{t=0}^{m} B_{s,n}(u)B_{t,m}(v)w_{s,t}}$$

The NURBS formula is the following [1]:

$$S(u,v) = \sum_{i=0,n} \sum_{j=0,m} R_{i,j}(u,v)P_{i,j} \tag{6.26}$$

NURBS is made for geometric design since getting the value of the weights w_{ij} is a huge problem. If we use a uniform rational and the weights w_{ij}'s are the same, then NURBS will be similar to B-spline.

6.4 Remarks: Finite Sample Points, Continuous Functions and Their Approximation

In theory, the Weierstrass approximation theorem states: For any continuous function, there must be a polynomial that approaches it. This means that if we can get a continuous function from a set of sample points, then we can easily get a smooth function. Therefore, fitting a set of sample points to a continuous function is essential.

It is possible to get a continuous function based on finite samples and then get the approximate polynomial. In fact, the Bernstein polynomial is such an implementation of the Weierstrass approximation theorem: Let $f(x)$ be continuous on $[0,1]$. If we make the Bernstein polynomial based on $f(x)$:

$$B_n(f)(x) = \sum_{i=0}^{n} f\left(\frac{i}{n}\right) B_{i,n}(x),$$

we can show that

$$\lim_{n \to \infty} B_n(f)(x) = f(x).$$

References

1. Cohen E, Riesenfeld R, Elber G (2001) Geometric Modeling with Splines. An introduction. Peters, Wellesley
2. Cormen TH, Leiserson CE, Rivest RL (1993) Introduction to algorithms. MIT, Cambridge
3. Kincaid D, Cheney W (2001) Numberical analysis, 3rd edn. Brooks-Cole, California
4. Preparata FP, Shamos MI (1985) Computational geometry: an introduction. Springer, New York
5. Press WH, Teukolsky SA, Vetterling WT, Flannery BP (2007) Numerical recipes: the art of scientific computing, 3rd edn. Cambridge University Press, New York
6. Quarteroni A (2009) Numerical models for differential problems. Springer, Milan/New York
7. Stoer J, Bulirsch R (2002) Introduction to numerical analysis, 3rd edn. Springer, New York

Chapter 7
Digital-Discrete Approaches for Smooth Functions

Abstract In this chapter, we present a systematic digital-discrete method for obtaining continuous functions with smoothness of a certain order (C^n) from randomly arranged data points. The new method is based on gradually varied functions and the classical finite difference method. This method is independent from existing popular methods such as the cubic spline method and the finite element method. The new digital-discrete method has considerable advantages for a large number of real data applications. This digital method also differs from other classical discrete methods that usually use triangulation.

7.1 Real World Needs: Looking for an Appropriate Fitting Method

We presented various mathematical approaches in Chap. 6 for function reconstruction, especially curve and surface fitting. For real applications, we mostly need to deal with surface fitting. There are two types of solutions for surface fitting: (1) For obtaining a continuous function, one can use piecewise linear functions to interpolate the data. The most popular and practical method is Delaunay triangulation. (2) Coons surface fitting, Bezier surface, and B-spline methods can construct smooth functions. However, they require that the sample points are selected along the boundaries of the rectangle cell or area. This technique is usually based on the tensor product of two curve fittings [4].

Some other methods can also apply to specific problems including finite element methods for partial differential equations, the subdivision method for computer graphics, and the moving least squares method for relatively dense sample data [17]. We explore these methods in Chaps. 11 and 12.

There are so many successful methods, so why do we still need a new method for smooth function reconstruction?

L.M. Chen, *Digital Functions and Data Reconstruction: Digital-Discrete Methods*,
DOI 10.1007/978-1-4614-5638-4_7, © Springer Science+Business Media, LLC 2013

This is because no method has achieved complete success in solving the following problem, a simple problem, generally speaking: Given a random set of sample points in 2D, how do we reconstruct a C^k function that fits the sample points?

7.1.1 Necessity of a New Method

In Chap. 3, we introduced gradually varied interpolation. The following simple example explains the necessity for developing such a method and introduces a new nonlinear method for smooth fitting.

		1	
4	[?]	[?]	3
		1	

This special example tries to fit the ? locations with reasonable numbers. Using triangulation, if the horizontal line is selected, we will have

		1	
4	[4-]	[3+]	3
		1	

If the vertical line is selected, we will have

		1	
4	[2+]	[1]	3
		1	

However, we really want

		1	
4	3	2	3
		1	

which is a solution of a gradually varied function.

The question now becomes whether any other method can help us achieve a similar result.

The finite elements method is one choice. However, it needs an initial triangulation. Then, the constraints of the boundary equation, possibly a partial differential equation, can be included using the basis function as it is in the B-spline method discussed in Chap. 6.

However, such a simple task requires a much more complex process. A benefit of using the finite elements method is the smoothness of the fitted function. Gradually varied functions in Chap. 3 could not produce a smooth result. We need to further develop a smooth function reconstruction method on top of using gradually varied functions.

7.1.2 Formal Description of the Reconstruction Problem

In this chapter, we will deal with the following application problems: (1) Given a set of points and the observation (function) values at these points, extend the values to a larger set. The reconstructed data will have the smoothness as described beforehand. (2) When observing an image, if we extract an object from the image, a representation of the object can sometimes be described by its boundary curve. If all values on the boundary are the same, then we can restore the object by filling the region. If the values on the boundary are not the same and if we assume the values are "continuous" on the boundary, then we need a fitting algorithm to find a surface.

Mathematically, the extension problem in this chapter can be defined as: Let D be a domain and J be a subset of D. If f is "continuous" or "smooth" on J, then is there a general method that yields an extension F of f for a set D with continuity or smoothness?

In continuous space, this problem is related to the Dirichlet problem if J is the boundary and the Whitney's problem if J is a subspace of D, where D is just the space R^n [13, 24]. The solution to the Dirichlet Problem is harmonic functions. We will discuss this in Chap. 9.

When J is randomly defined, this general extension problem still attracts many research scientists and engineers. This chapter provides a totally new approach to solving this problem.

Why are existing numerical methods not perfect? If the boundary is irregular, then we need to use a 2D B-spline to divide the boundary into four segments where different partitions yield different results. We will do a comprehensive review in Chap. 8.

The main concern is: What function can be called "continuous" or "smooth" in discrete space? In Chap. 3, we defined gradually varied functions that can be referred to as continuous functions in discrete space [4].

Also in Chap. 3, we presented the necessary and sufficient conditions for the existence of a gradually varied extension F : for all x,y in J, $d(x,y) \leq |i - j|$ where $f(x) = A_i$, $f(y) = A_j$, and d is the distance between x and y in D.

What about smoothness? Can we extent the gradually varied function to smooth gradually varied functions? Also, how would we define them? We will address these issues through the finite difference method in the next section.

7.2 Smooth Gradually Varied Functions

The key to the new method for reconstructing a smooth gradually varied function is to first calculate a "continuous" (i.e. gradually varied) function, then obtain the partial derivatives, and finally modify the original function (to be smoother).

A method to keep the gradually varied property of the partial derivative function is designed to fulfill necessary part of the definition. This procedure can be recursively done in order to get high order derivatives. Then, we can use Taylor expansion for the local fitting. The use of Taylor expansion for a 2D surface was designed by many researchers [9, 14, 18].

Even though the description of the method appears to be hard to follow, the actual process can be explained step-by-step. There should not be any technical difficulties.

7.2.1 Derivatives of Gradually Varied Functions

This section deals with how to calculate the derivatives using a fitted gradually varied function.

Given $J \subseteq D$, and $f_J : J \to \{A_1, A_2, \cdots, A_n\}$. Let f_D be a gradually varied extension of f_J on D, which is a simply connected region of 2D grid space. According to Sect. 2.4, the forward difference equation for derivatives with $\Delta_x = \Delta_y = 1$ is

$$f_D'|_x = \frac{\partial}{\partial x} f_D(x,y) = f_D(x+1,y) - f_D(x,y) \tag{7.1}$$

$$f_D'|_y = \frac{\partial}{\partial y} f_D(x,y) = f_D(x,y+1) - f_D(x,y) \tag{7.2}$$

For better error control, we may select the central difference formulas

$$f_D'|_x = \frac{\partial}{\partial x} f_D(x,y) = f_D(x+0.5,y) + f_D(x-0.5,y) - 2f_D(x,y), \tag{7.3}$$

$$f_D'|_y = \frac{\partial}{\partial y} f_D(x,y) = f_D(x,y+0.5) + f_D(x,y-0.5) - 2f_D(x,y). \tag{7.4}$$

Or use some estimated formulas. The general difference method and calculations of partial derivatives was introduced in Chap. 2.

These derivatives will be regarded as the estimation for the fitted surfaces. Sometimes we knew the derivative values for the whole or subset of the domain. Then we will use these instead of the above equations. Sometimes, if we are not confident with the whole function (e.g. the function was calculated by the above equations), we will use gradually varied functions to fit the samples of derivative values.

With the new method, the digital-discrete method [7, 9], after we have the derivatives f' then we can use them to re-calculate or update the original (gradually varied) fitted function by adding the first derivative components.

This method can be used to calculate all different orders of derivatives. The mixed mechanism can then be employed: calculate derivatives based on the gradually varied fitting, then complete a gradually varied fitting on the derivatives; repeat the previous two steps.. In addition, the Taylor expansion at the sampling point will be applied to a region with a certain radius (to increase the accuracy of the Taylor expansion).

7.2.2 Definition of Gradually Varied Derivatives

In Lemma 3.1 of Sect. 3.7, we have proven that any continuous function f, in constructive compact metric space M, has a gradually varied approximation. Therefore, we have:

Lemma 7.1. *If $f^{(k)}$, the k-order of derivatives, is continuous, then $f^{(k)}$ is digitally continuous (or gradually) varied in a digitization of M.*

The basic consideration of this chapter is to find a smooth function for discrete space. Not only must the function be continuous, but it must also contain continuous derivatives of a certain degree C^n. In 2010, Chen invented the concept of gradually varied derivatives [9]. This is the central concept in this chapter.

Definition 7.1. Let f be a gradually varied function on Σ_m. If the derivative of f calculated by the finite difference method is gradually varied, then we say that f has gradually varied derivatives.

Discrete smooth means the same as gradually varied derivatives in that the derivatives are also gradually varied. Using gradually varied functions to actually obtain gradually varied derivatives will be explored in next subsection.

One of the main purposes of this new method is to obtain the "continuous" derivative function using gradually varied reconstruction.

After a function is reconstructed, one can get all orders of the derivatives. However, it is not guaranteed that we will have "smooth" derivatives. In order to maintain the continuity of derivative functions, we either need to smooth the derivative function or use another method to make a continuous function. In some situations, we can get the values of not only the function itself but also the value(s) of the derivative function.

For surface reconstruction, we directly use gradually varied fitting to get the gradually varied functions f_x and f_y. Then we can use the same technique to get f_{xx}, f_{xy}, and f_{yy}. Eventually, we can get as many orders of derivatives as we would like.

Definition 7.2. $GVF(f_J)$ is defined as a gradually varied interpolation of f_J on subset J in D. $f^{(0)}$ denotes the "continuous" interpolation of f_J on D, so $GVF(f_J)$ is used as $f^{(0)}$. (Or simply $f^{(0)} = GVF(f_J)$).

For simplification, denote $G(f) = GVF(f)$. $G(f)$ could have two meanings: (1) f is already on D, G(f) is a gradually varied uniform approximation of f. (2) $G(f)$ is gradually varied interpolation on a subset $S \subset D$. We usually choose $S = J$ or little larger than J, i.e. $S \supseteq J$. We sometimes use $G_S(f)$ to represent this special gradually varied interpolation.

Let $g = f^{(0)} = G(f_J)$. $g' \approx \frac{\Delta g}{\Delta x}$. Since $\Delta x = 1$, so $g' \approx \Delta g$. We can just use $g' = (G(f_J))'$ for simplification. Thus, $f' = (g)' = (f^{(0)})'$, so

$$f^{(1)} = G(g') = G((f^{(0)})') = G(G'(f_J))$$

Define
$G^{(k)}(f^{(0)}) = G((G^{k-1}(f^0))') = G(G^{k-1}(f^0))'.$
$G^{(k)}(f^{(0)})$ is an approximation of $f^{(k)}$, the kth-order of derivatives that is also continuous (the kth-order continuous derivatives).

For example, $(G(f'))' = f''$, but f'' is not $f^{(2)}$ since f'' may not be continuous. The statement is $f^{(2)} = G(f'')$. In addition, $f'' = \frac{\Delta^2 f}{\Delta x^2}$ is no longer valid since f' might not be continuous. The meaning of f'' is really $(G(f'))'$. In summary, the kth-order gradually varied derivatives is

$$f^{(k)} = G^{(k)}(G(f_J)) = G^{(k)}(f^{(0)}) = G((G^{k-1}(f^0))'), \qquad (7.5)$$

where $G^{(0)}(G(f_J)) = G(f_J) = f^{(0)}$. This formula presents an alternative process on calculating the approximation of derivatives: (1) Use gradually varied functions to get the surface, (2) use the finite method to get the derivatives. This is a key conceptual discovery in [9].

For some applications, we may know $f^{(1)}$, but we want to get $f^{(3)}$. Assume that f's k'-order continuous derivatives are known. If we want to get a $K = k + k'$-order derivatives, we can also define

$$G^{(k)}(f^{(k')}) \approx f^{(K)}. \tag{7.6}$$

This is the idea of gradually varied derivatives.

Some other investigations are interesting. Assume f is (naturally) gradually varied (or digitally continuous), but f' may not be gradually varied. Let $G_\alpha(f')$ be a gradually varied approximation, we define

$$|G_\alpha(f') - f| = \max\{|G_\alpha(f')(p) - f'(p)| : p \in D\} \tag{7.7}$$

Studying the property of $G_{opt}(f')$ is an important problem, where

$$|G_{opt}(f') - f'| = \min\{|G_\alpha(f') - f'| :$$
$$G_\alpha(f') \text{ is a gradually varied approximation of } f'. \} \tag{7.8}$$

Since J is the original subset where f' may exist, we usually use a $J' \supseteq J$ to get $G(f')$. The algorithm will be presented in later sections.

Open Problem 7.1. Estimation of $h = |G_{opt}(f') - f'|$ and further $h_i = |G_{opt}(f^i) - f^i|$. Study if there exists a convergence theorem such that

$$\lim_{n \to \infty} h_i = 0? \tag{7.9}$$

Or, in what condition does, $\lim_{n \to \infty} h_i = 0$.

7.2.3 Recalculation of the Function Using Taylor Expansion

In calculus, the Taylor expansion theorem is a central theorem that states: A differentiable function f around a point x_0 can be represented by a polynomial composed by the derivatives at the given point x_0. The one variable formula is as follows:

$$f(x) = f(x_0) + f'(x_0)(x - x_0) + \frac{f''(x_0)}{2!}(x - x_0)^2 + \cdots +$$
$$\frac{f^{(k)}(x_0)}{k!}(x - x_0)^k + R_k(x)(x - x_0)^k, \tag{7.10}$$

where $\lim_{x \to x_0} R_k(x) = 0$. The polynomial in (7.10) is called the Taylor polynomial or Taylor series. $R_k(x)$ is called the residual. This theorem provides us with a theoretical foundation of finding the k-th order derivatives at point x_0 since we can restore the functions around x_0.

After the different derivatives are obtained, we can use Taylor expansion to update the value of the gradually varied fitted function (at C^0). In fact, in any order C^k, we can update the function using a higher order of derivatives as discussed in the above section. For an m-dimensional space, the Taylor expansion has the following generalized form expanding at the point $(a_1, \ldots, a_k)$:

$$f(x_1, \ldots, x_k) = \sum_{n_1=0}^{\infty} \cdots \sum_{n_k=0}^{\infty} \frac{(x_1 - a_1)^{n_1} \cdots (x_k - a_k)^{n_k}}{n_1! \cdots n_k!} \left(\frac{\partial^{n_1 + \cdots + n_k} f}{\partial x_1^{n_1} \cdots \partial x_k^{n_k}} \right) (a_1, \ldots, a_k).$$

(7.11)

For a function with two variables, x and y, the Taylor series of the second order at expanding point (x_0, y_0) is:

$$f(x, y) \approx f(x_0, y_0) + (x - x_0) \cdot f_x(x_0, y_0) + (y - y_0) \cdot f_y(x_0, y_0) +$$
$$\frac{1}{2!} \left[(x - x_0)^2 \cdot f_{xx}(x_0, y_0) + 2(x - x_0)(y - y_0) \cdot f_{xy}(x_0, y_0) + \right.$$
$$\left. (y - y_0)^2 \cdot f_{yy}(x_0, y_0), \right]$$

(7.12)

There are several ways to implement this formula. For smooth gradually varied surface applications, $f(x_0, y_0)$, f_x, and f_y are $G(f)$, $G^{(1)}(f)$, etc.

In general, based on (7.5), we calculate the following gradually varied derivatives: Proposition 7.1

(a) $G = GVF(f_I)$,
(b) $G_x = GVF(\frac{\partial G}{\partial x})$,
(c) $G_y = GVF(\frac{\Delta G}{\Delta y})$,
(d) $G_{xx} = GVF(\frac{\Delta G_x}{\Delta x})$,
(e) $G_{xy} = GVF(\frac{\Delta G_x}{\Delta y})$, and
(f) $G_{yy} = GVF(\frac{\Delta G_y}{\Delta y})$.

By (7.12), we will have

$$f(x, y) \approx G(x_0, y_0) + (x - x_0) \cdot G_x(x_0, y_0) + (y - y_0) \cdot G_y(x_0, y_0) +$$
$$\frac{1}{2!} \left[(x - x_0)^2 \cdot G_{xx}(x_0, y_0) + 2(x - x_0)(y - y_0) \cdot G_{xy}(x_0, y_0) + \right.$$
$$\left. (y - y_0)^2 \cdot G_{yy}(x_0, y_0), \right]$$

(7.13)

The above formula shows the principle of digital-discrete reconstruction. Once the derivatives of sample points are obtained, the Taylor expansion formula can be used to get the values around the sample points. It is not appropriate to obtain values far from the sample points since if the residual R_k (7.10) is large enough, it may generate large errors.

A simple method can be designed to simply fill this need. Starting at the sample point, do an expansion based on its neighbors and then do another expansion. Once we reach the points adjacent to more than one point with an updated value, average these values using Taylor expansion.

This algorithm will result in k-smooth functions, since it is a polynomial or a linear combination of polynomials. Another algorithm uses the neighboring points to reconstruct the center point. To get the interpolation, we do not update the sample points. For an approximation, we could update the sample points.

In this approach, the algorithm that takes 0-order parts of the Taylor series will perform an average solution, just like data filtering in image processing. This algorithm can also refer to as Taylor jets that have connections to the moving least squares method.

The major difference between our method and existing methods is that we designed a way to calculate derivatives at the sample points using the digital-discrete method. This method does not rely on the shape of the domain and locations of sample points. Again, using gradually varied fitting to fit the surface, we then restore the value to calculate the derivatives. Using the Taylor expansion to cover the area, this method is similar to a patch or a jet—a function on small local area , a Taylor expansion on a patch. The Taylor expansion formula can also be changed into Bezier or B-spline. We discuss this in Chap. 11.

In Sect. 7.3, we discuss the detailed algorithms. In practice, we choose a ratio that changes less than half the function's value. An iteration process is designed to make the new function convergent.

7.3 Algorithm Design and Digital-Discrete Methods

In this section, we discuss algorithm design issues. In Sect. 7.2, a systematic digital-discrete method for obtaining continuous functions with smoothness of a certain order (C^n) from the sample data was introduced. In order to implement this method in this section, we design algorithms to accomplish our task.

7.3.1 The Main Algorithm

The new algorithm tries to search for the best solution to the fitting. We have added a component of the classical finite difference method. The major steps of the new algorithm include (for 2D functions only, for 3D functions we only need to add one dimension):

Algorithm 7.1 (The main procedure of the implementation:).

Step 1: Load guiding points. In this step we load the data points with observation values.

Step 2: Determine resolution. Locate the points in grid space.

Step 3: Function extension according to the theorem presented in Sect. 3.2 (Algorithm 3.1). This way, we obtain gradually varied or near gradually varied (continuous) functions. In this step, the local Lipschitz condition might be used.

Step 4: Use the finite difference method to calculate partial derivatives. In this step, we restore actual values for original sample points. When the grid is not dense, we may need to restore the actual location too. Then obtain the derivatives.

Step 5: Use Taylor expansion to obtain the smoothed function.

Step 6: A multilevel or multiple resolution method may be used.

7.3.2 The Algorithm for the Best Fit

When the guiding points do not satisfy the condition of gradual variation, an algorithm for the best possible fit is designed. This algorithm will leave some gaps in the surface that may not be gradually varied. In our software, we have implemented such an algorithm.

This algorithm is similar to Algorithm 3.1. However, we will not stop to produce a surface when the gradually varied condition ($LD(p,p) \le d(p,p'), p, p' \in J$) is not satisfied. For a new point x assigned with the value of its neighbor p's value $f(p)$, we calculate $LD(x,p') - d(x,p')$, where $p' \in J$. If $LD(x,p') - d(x,p')$ is positive, the assigned values will not be gradually varied. There are two standards we can follow: (1) Select $f(x)$, such at

$$\min \sum_{p' \in J} \{LD(x,p') - d(x,p')| \text{ if } LD(x,p') - d(x,p') > 0\}$$

or (2) Select $f(x)$, such at

$$\min\{ \text{ the number of } LD(x,p') - d(x,p') > 0, p' \in J\}.$$

We usually set $f(x) = f(p)$, but this is not necessary.

Algorithm 7.2 (Best Fit). Let f_J be a function defined on a non-empty subset J of D. Let $D_0 = J$:

Step 1: Choose x from $D - D_0$ such that x has an adjacent vertex r in D_0, assume $f(r) = A_i$ and $f(x) = A_i$.

Step 2: Select $f(x)$ from A_{i-1}, A_i, and A_{i+1} such that the number of $LD(x,p') - d(x,p') > 0$, where $p' \in D_0$ is the minimum.

Step 3: Let $D_0 \leftarrow D_0 \cup \{x\}$.

Step 4: Repeat Steps 2–4 until $D_0 = D$.

7.3.3 The Algorithm for the Gradually Varied Uniform Approximation

Another algorithm for making a gradually varied surface when the guiding points do not satisfy the condition of gradual variation uses the uniform approximation [3, 4].

Algorithm 7.3 (Chen 1994). The algorithm for uniform approximation can be designed as follows:

Step 1: Initially, let $i \leftarrow 1$, and compute $F_i(p)$, $F_i(p)/F_i(q)$, and $F_i^k(p)$.

Step 2: The condition of Theorem 11 must be satisfied. If it does not hold, then $i \leftarrow i + 1$ and go back to Step 1.

Step 3: Let $g_J(p) = \inf\{F_i^k(p)\}$ for each $p \in J$.

Step 4: Get a gradually varied extension f_D of g_J.

A proof of the correctness of this algorithm can be found in [4].

7.3.4 The Iterated Method for Calculating Derivatives

In this section, we present more details for calculation of gradually varied derivatives. After we have obtained f_x and f_y from $F = GVF(f_J)$, we can then re-compute the gradually varied fitting based on $f_x = F'|_x$ and $f_y = F'|_y$.

First, get a smooth f_x and f_y by applying a gradually varied fitting on f_x and f_y, so we have $F_x = GVF(f_x)$ and $F_y = GVF(f_y)$ based on J. By Taylor expansion, use F, F_x, and F_y to get a $C^{(1)}$ function, F_{new}. Here is the example for F_x and F_y in Fig. 7.1:

We need to update the results until the new function can no longer be changed. Then we have a fixed f_x or f_y. So we can do gradually varied fittings on the selected points in f_x or f_y, then repeat. We will get f_{xx}, f_{xy}, and f_{yy}. Update f_x (or f_y), until it can no longer be changed or the difference of the change is limited in a given range. Then we return back to change f. If we know f_x or f_y, we can use f_x and f_y to guide the fitting directly.

In other words, this method uses F_x and F_y, or $GVF(f_x)$ and $GVF(f_y)$, to update $F = GVF(f)$. Forgetting the scale A's and using numerical updates, we use iteration based on the fitting orders.

Two sub-methods can be chosen in this consideration. Update the whole F, and then compute F_x and F_y. Then, repeat the process until no change or changes smaller than the threshold assigned. Update the function based on each order (distance to the guiding points) and then redo F_x and F_y, using the updated points as guiding points. Repeat.

Until no further changes are possible, we get F_x, F_y, or $GVF(f_x*)$, $GVF(f_y*)$ (* where we add extreme points, i.e. choose $J*$ where $J \subseteq J*$). Then we compute F_{xx}, F_{xy}, and F_{yy}, and so on and so forth. Using F_{xx}, etc. to update F_x and F_y, we then go back to update F again. This is the procedure.

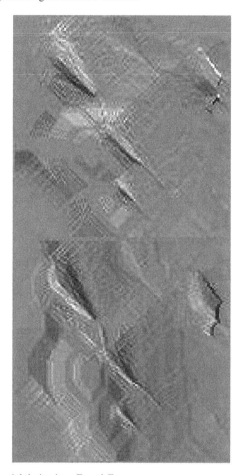

Fig. 7.1 First order partial derivatives F_x and F_y

7.3.5 The Multi-scaling Method

The multi-scaling method is used to choose a base scale and then refine the scale by 2 (dividing the edge length by 2). This is just like the wavelet method. Other multi-scaling methods for partial differential equations can be found in [25].

If there is more than one guiding point in a pixel or block unit, we can chose one or use their average value. Computing the gradually varied function at scale k (usually, the length of the rectangle cell is $(length \cdot 2^{-k})$, then $F_k = GVF(f,k)$, we can get $F_{\{x,k\}}$. We then calculate and insert the value at 1/2 point intervals surrounding the guiding points (restore the original sample points when the smaller scale is placed).

Compute the whole insertion or do it in order before using the new calculated points as guiding points for gradually varied functions. This gradually varied function is on the new scale. This will guarantee that the derivatives are guiding points.

Then we use this to refine the scale again by keeping the inserted points as guiding points. We will have more points surrounding the original guiding points, and so on. We can get our F in a predefined scale, which will also help get a good derivative.

Using the subdivision method, we can also make up to an almost C^2 function. We discuss this method in Chap. 12.

We can also use this to calculate F_{xx}, F_{xy}, and F_{yy}, based on the new gradually varied function to refine F. The export will be smoothed at the order we choose.

7.3.6 Case Study: Reconstruction Experiments

A practical method will have some real examples to support the results. Three sets of real data are tested. The second set uses the same ten sample points (Fig. 7.2)

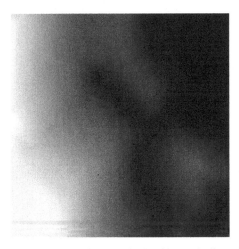

Fig. 7.2 Northern VA Groundwater distribution calculated by gradually varied surfaces date from 04/01/07. Ten sample points are used at Day 95

A set of data contains 29 sample points. Using the algorithms described in the above sections, we have calculated all types of derivatives and made function reconstruction for C^0, C^1, and C^2 functions. Pictures are shown in Fig. 7.3. One can see that Fig. 7.3c contains a much smoother image than that of Fig. 7.3a.

7.4 Derivative Calculation Using Global Methods

The above finite difference method for derivative calculation can be viewed as a local method for derivatives. This subsection designs a global method using Bernstein polynomial approximation. In terms of smooth functions based on the Bernstein polynomial approximation and Bezier or Bspline methods, we will discuss these in Chap. 12.

The methodology used in this section is that once we have obtained the initial GVF reconstruction, we apply a Bernstein polynomial approximation in the x-axis or y-axis. The sample points for this polynomial can all be selected or a part randomly selected as GVF fitted points. Then, we have the polynomial and can find mathematical derivatives for each grid point in the x-axis to form F_x, so we can get F_y just as we did in last section. Then, these are combined.

The philosophy behind this is to use the global information of the entire surface to predict the derivatives at a point. The local details will be obtained after the iteration. The Bernstein polynomial is a uniform approximation for a continuous function, it performs particularly well when dense sample points are provided.

GVS gives such an initial continuous function. One could also use a piece-wise linear function such as triangulation.

According to the approximation formula in Chap. 6, If f is a continuous function on the interval $[0,1]$, then f_i is the value at i/n. The Bernstein polynomial:

$$P(x) = \sum_{i=0}^{n} f_i b_{i,n}(x) = \sum_{i=0}^{n} f_i \binom{n}{i} x^i (1-x)^{n-i} \tag{7.14}$$

$$P'(x) = \sum_{i=0}^{n} f_i b'_{i,n}(x) = \sum_{i=0}^{n} f_i \binom{n}{i} \left(i x^{i-1}(1-x)^{n-i} - (n-i)x^i(1-x)^{n-i-1} \right) \tag{7.15}$$

In fact, for easier calculation, the derivative can be written in base polynomials of a lower degree:

$$b'_{i,n}(x) = n \left(b_{i-1,n-1}(x) - b_{i,n-1}(x) \right). \tag{7.16}$$

Bernstein polynomials are a type of implementation to the Weierstrass theorem: every real-valued continuous function on a real interval $[a,b]$ can be uniformly approximated by polynomial functions over R. A more explicit form of Bernstein polynomials is

$$b_{i,n}(x) = \binom{n}{i} x^i (1-x)^{n-i} = \sum_{k=i}^{n} (-1)^{k-i} \binom{n}{k}\binom{k}{i} x^k \tag{7.17}$$

$$P(x) = \sum_{i=0}^{n} \sum_{k=i}^{n} (-1)^{k-i} \binom{n}{k}\binom{k}{i} f_i \cdot x^k \tag{7.18}$$

$$P'(x) = \sum_{i=0}^{n} \sum_{k=i}^{n} (-1)^{k-i} \binom{n}{k}\binom{k}{i} k \cdot f_i \cdot x^{k-1} \tag{7.19}$$

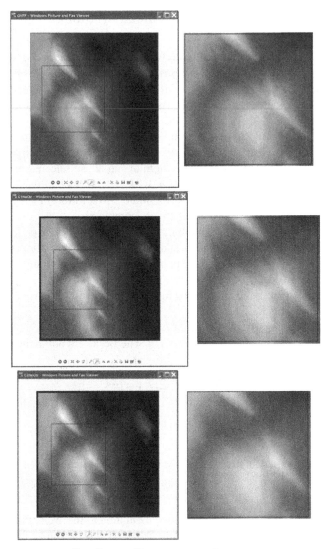

Fig. 7.3 Comparison for $C^{(0)}$, $C^{(1)}$ and $C^{(2)}$

For computer calculations in coding and programming, recursive formulas are much more time efficient. As we mentioned before, in order to make the calculation viable, the resolution of samples in the GVF fitted surface can also vary. We can even make select local Bernstein polynomials, say 3–4th order, to get the local derivatives. Figure 7.2 shows the $C^{(2)}$ function using Bernstein polynomials for calculating f_x and f_y, etc. Compared to the gradually varied iterated method, the image boundary is not as clear as the image in Fig. 7.1. It seems like the global method has averaged the information in the data.

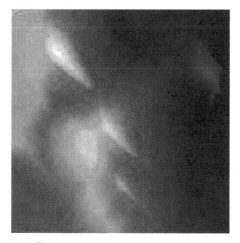

Fig. 7.4 The reconstructed $C^{(2)}$ function using Bernstein polynomials

It is true that there is a direct way of using Bernstein polynomials in lower degrees for direct fitting without calculating the derivative. This method is related to Bezier and B-spline methods, which we discuss in Chap. 12 [12, 20].

7.5 Relationships to Functional Analysis Methods: Lipschitz Extensions

Extension of Lipschitz functions in functional analysis has been studied for almost 80 years, beginning in 1934 with McShane, Whitney, and Kirszbraun's seminal papers [19, 23, 24]. This research was essential in theory [15]. This problem is also related to Whitney's problem [14, 16, 21]: Given a function f on a subset of a space (Euclidean or Sobolev space), can we find an extension of f that has the same smoothness as f?

7.5.1 McShane-Whitney Extension Theorem

The McShane-Whitney extension theorem states that a Lipschitz function f on a subset J of a connected set D in a metric space can be extended to a Lipschitz function F on D. McShane gives a constructive proof for the existence of the extension in [19].

Theorem 7.1. *Assume f is a L-Lipschitz function, i.e. $|f(a) - f(b)| \leq L|a - b|$ for all a, b in J. Then $F(x) = \min_{a \in J}\{f(a) + L|x - a|\}$ is a L-Lipschitz extension of f on D;*

Proof. The McShane-Whitney construction is very elegant [15, 19]. We will provide an easy (elementary) proof here. Let $|f(a) - f(b)| \leq L|a - b|$ for all a, b in J.

$$f_a(x) = f(a) + L|x - a|$$

so $f_a(a) = f(a)$.

First, we will see that $f_a(x)$ is also L-Lipschitz for the entire space since

$$|f_a(x) - f_a(y)| = L|(|x - a| - |y - a|)|$$

We can also see that $|(|x - a| - |y - a|)| \leq |x - y|$. This is based on the triangle inequality, which is always true in any metric space $|x - a| \leq |y - a| + |x - y|$. Thus, $|f_a(x) - f_a(y)| = L|x - y|$.

Define $F(x) = \min_{a \in J}\{f_a(x)\}$. For any $x, y \in D$, there exists a and b such that $F(x) = f(a) + L|x - a|$; $F(y) = f(b) + L|y - b|$.

If $F(x) \leq F(y)$, we know $F(y) = f(b) + L|y - b| \leq f(a) + L|y - a|$ $0 \leq F(y) - F(x) \leq f(a) + L|y - a| - f(a) - L|x - a| \leq L(|y - a| - |x - a|)$

So $|F(y) - F(x)| \leq L|(|y - a| - |x - a|)|$. As discussed above, $|F(y) - F(x)| \leq L|y - x|$.

If $F(x) > F(y)$, then $F(x) = f(a) + L|x - a| \leq f(b) + L|x - b|$, $0 \leq F(x) - F(y) \leq f(b) + L|x - b| - f(b) - L|y - b|$. Therefore, $|F(x) - F(y)| \leq L|x - y|$.   □

7.5.2 Comparison Between Gradually Varied Functions and Lipschitz Extension

McShane-Whitney's construction $F(x) = \min_{a \in J}\{f(a) + L|x - a|\}$ is a maximum extension. We can let $F'(x) = \min_{a \in J}\{f(a) - L|x - a|\}$ be a minimum extension. It is obvious that neither $F(x)$ nor $F'(x)$ can directly be used in data reconstruction. However, $H = (F + F')/2$ is a reasonable function that maintains good properties including each value at a point contained in the convex of guiding points. Also, H is a Lipschitz extension. For the purposes of this chapter, let's call H the McShane-Whitney mid function. The result of applying this function to our data sets is shown below.

We can see that Fig. 7.5a is not a good fit compared to Fig. 7.2. Figure 7.5b is a reasonable reconstruction compared to Figs. 7.3 and 7.4. The fitting is dominated by the Lipschitz constant, so the extension function is "controlled" by sample points with bigger values.

Another factor is caused by the algorithmic difference. The difference between our algorithm and the McShane-Whitney construction method is more than the theoretical difference. The digital-discrete algorithm design based on the proof of Theorem 3.1 in Chap. 3 is discrete and dynamic. The digital-discrete method also adjusts locally. The McShane-Whitney extension method seems to be mainly for theoreti-

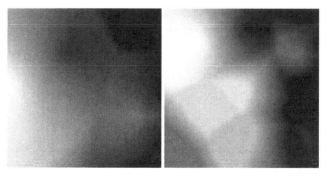

Fig. 7.5 McShane-Whitney mid extensions: (**a**) Using the sample data set of Fig. 7.2, (**b**) Using the sample data set of Figs. 7.3 and 7.4

cal purposes. It is simple and may be fast in calculation. However, it loses flexibility especially since it is just for the Lipschitz function.

7.6 Remark

To get a smoothed function using gradual variation is a long time goal of our research [1–5, 7]. Some theoretical attempts have been made before, but struggled in the actual implementation [10, 11]. The author was invited to give a talk at the Workshop on the Whitney's Problem organized by Princeton University and the College of William and Mary in August 2009. He was in part inspired and encouraged by the presentations and the helpful discussions with the attendees at the workshop. We finally have made the concept of smooth gradually varied function to solve this problem through the development of the concept of gradually varied derivatives [6–9]. It takes more than twenty years to come up with the idea of gradually varied derivatives.

Any continuous function can be approximated by a sequence of gradually varied functions (Chap. 3). An analytic function is also a continuous function. So it is done by just using the property. Use $G^{(k)}$ to simulate $f^{(k)}$ is necessary since the finite difference in high order may not valid when the lower order derivatives not continuous.

The fundamentals of this method differs from that of Fefferman's theoretical method, which uses a system of linear inequalities and an objective function (called linear programming) to find the solution at each point. The inequalities are for all different orders of derivatives.

Acknowledgements This research has been partially supported by the USGS Seed Grants through the UDC Water Resources Research Institute (WRRI) and Center for Discrete Mathematics and Theoretical Computer Science (DIMACS) at Rutgers University. Professor Feng Luo suggested the direction of the relationship between harmonic functions and gradually varied functions. Dr. Yong

Liu provided many helps in PDE. UDC undergraduate Travis Branham extracted the application data from the USGS database. Professor Thomas Funkhouser provided helps on the 3D data sets and OpenGL display programs. The author would also like to thank Professor C. Fefferman and Professor N. Zobin for their invitation to the Workshop on the Whitney's Problem in 2009.

References

1. Chen L (1990) The necessary and sufficient condition and the efficient algorithms for gradually varied fill. Chinese Sci Bull 35(10):870–873
2. Chen L (1992) Random gradually varied surface fitting. Chinese Sci Bull 37(16):1325–1329
3. Chen L (1994) Gradually varied surface and its optimal uniform approximation. In: *IS&T* SPIE symposium on electronic imaging, SPIE Proceedings, San Jose, Vol 2182
4. Chen L (2004) Discrete surfaces and manifolds. Scientific and practical computing. Rockville, Maryland
5. Chen L (2005) Gradually varied surfaces and gradually varied functions, in Chinese, 1990; in English 2005 CITR-TR 156, University of Auckland
6. Chen L, Applications of the digital-discrete method in smooth-continuous data reconstruction. http://arxiv.org/ftp/arxiv/papers/1002/1002.2367.pdf
7. Chen L, Digital-discrete surface reconstruction: a true universal and nonlinear method. http://arxiv.org/ftp/arxiv/papers/1003/1003.2242.pdf
8. Chen L (2009) Gradual variation analysis for groundwater flow of DC (revised), DC Water Resources Research Institute Final Report, Dec 2009. http://arxiv.org/ftp/arxiv/papers/1001/1001.3190.pdf
9. Chen L (2010) A digital-discrete method for smooth-continuous data reconstruction. J Wash Acad Sci 96(2):47–65. (ISSN 0043-0439), http://arxiv.org/ftp/arxiv/papers/1010/1010.3299.pdf
10. Chen L, Adjei O (2004) lambda-connected segmentation and fitting. In: Proceedings of IEEE international conference on systems man and cybernetics, Orlando, vol 4, pp 3500–3506
11. Chen L, Liu Y, Luo F (2009) A note on gradually varied functions and harmonic functions. http://arxiv.org/PS_cache/arxiv/pdf/0910/0910.5040v1.pdf
12. Catmull E, Clark J (1978) Recursively generated B-spline surfaces on arbitrary topological meshes. Comput Aided Des 10(6):350–355
13. Courant R, Hilbert D (1989) Methods of mathematical physics, vol 1. Wiley, New York
14. Fefferman C (2009) Whitney's extension problems and interpolation of data. Bull Am Math Soc 46:207–220
15. Heinonen J (2005) Lectures on lipschitz analysis, report. Department of Mathematics and Statistics, vol 100, University of Jyvaskyla, Jyvaskyla, 2005
16. Klartag B, Zobin N (2007) C1 extensions of functions and stabilization of Glaeser refinements. Rev Math Iberoam 23(2):635–669
17. Lancaster P, Salkauskas K (1981) Surfaces generated by moving least squares methods. Math Comput 87:141–158
18. Mallet J-L (1989) Discrete smooth interpolation. ACM Trans Graph 8(2):121–144
19. McShane EJ (1934) Extension of range of functions. Bull Am Math Soc 40:837–842
20. Peters J (1993) Smooth free-form surfaces over irregular meshes generalizing quadratic splines. Comput Aided Geom Des 10(3–4):347–361
21. Shvartsman P (2009) On Sobolev extension domains in Rn. http://arxiv.org/abs/0904.0909
22. Thurston W (1997) Three-dimensional geometry and topology. Princeton University press, Princeton
23. Valentine FA (1945) A Lipschitz Condition Preserving Extension for a Vector Function. Am J Math 67(1):83–93
24. Whitney H (1934) Analytic extensions of functions defined in closed sets. Trans Am Math Soc 36:63–89
25. Yue X, Weinan E (2005) Numerical methods for multiscale transport equations and application to two-phase porous media flow. J Comput Phys 210(2):656–675

Chapter 8
Digital-Discrete Methods for Data Reconstruction

Abstract This chapter is the continuation of Chap. 7. It focuses on various applications and their data representations of the digital-discrete method. This chapter is divided into three parts: (1) An introduction to real problems and then a discussion of the data structure for the reconstruction problems. Then we focus on the implementation details. This part is specifically for real data processing for 2D and 3D domains. (2) Function reconstruction on manifolds. We will first introduce discrete manifolds (meshes) and digital manifolds, then discuss data reconstruction on manifolds. (3) At the end of the chapter, we discuss the methodology issues of our method and its differences from between other methods. Piecewise smooth data reconstruction and harmonic data reconstruction is presented in Chap. 9. Smooth data reconstruction on triangulated manifolds using the subdivision method is introduced in Chap. 12.

8.1 Real World Problems

Data fitting and reconstruction is always at the center of the computational sciences and engineering. It affects peoples' everyday lives ranging from medical imaging, bridge construction, and rocket launching to the auto industry, global warming, and weather forecasting. All of these use numerical methods including numerical partial differential equations for data reconstruction.

The general form of the problem is as follows: Based on finite observations, how do we extend a solution to whole area with continuity and smoothness?

We list some of the common problems here:

(a) When a suspect vehicle is driving in the city, several cameras captured the vehicle. The authority wants to know the driving path and predict where the vehicle will drive to next.
(b) When BP's Macondo well leaked oil in the Gulf of Mexico, the Virginia coast guard sent a ship to investigate whether Virginia is affected. The ship drives

L.M. Chen, *Digital Functions and Data Reconstruction: Digital-Discrete Methods*,   99
DOI 10.1007/978-1-4614-5638-4_8, © Springer Science+Business Media, LLC 2013

randomly and picks up some samples. If there are traces of oil, we want to know the entire distribution of the effects.

(c) When boundary values are known, find the integral data that fits the boundary.

(d) In a dry summer, residents pump ground water to irrigate the lawn from the wells. If they pump too much, the agricultural plants will lack water in rural areas outside the city. So, how much water can be pumped to water lawns?

(e) For the car industry, designing the shape of the car usually requires computer aided design (CAD) technology. A designer usually determines certain values at a few positions on the car body and a program in the CAD tool then automatically fits a surface lying on the designated points.

(f) If a train track runs on a bridge, engineers examine the pressure in some critical points. When data is brought to the lab, the computational engineer needs to map out the pressure on the entire bridge.

The above list gives a few examples of data reconstruction. Many effective methods have been developed to solve these problems. However, a general solution to these problems is to investigate how to get a smooth function to go through the sample points. In Chap. 7, we presented a digital-discrete solution.

In this chapter, we focus on how to apply the new method to several practical problems.

First, we describe the general process of applying the digital-discrete method. Then, we introduce the effective data structures for data reconstruction.

The classical discrete method for data reconstruction is based on domain decomposition according to guiding (or sample) points. Then, the spline method (for polynomial) or finite elements method (for PDE) is used to fit the data. Our method is based on gradually varied functions that do not assume the property of being linearly separable among guiding points, i.e. no domain decomposition methods are needed. We also demonstrate the flexibility of the new method and its potential in solving a variety of problems. The method can easily be extended to higher multi-dimensions. In this chapter, the experimental data sets come from a modified example in Princeton's 3D Benchmark data sets.

8.2 General Process for Digital-Discrete Methods

In Chap. 7, we presented the core theory and algorithms for the digital-discrete method. Here, we articulate the steps for a detailed implementation starting with the raw data and ending with a graphical display. This method contains the following major steps:

First, load the sample or guiding points. In this step we load the raw data points with observation values. The data format may be different, but the most common formats are as follows. Data can be listed as a set of vectors, such as $< p, v >$, where p is the point and v is the value. Both p and v can be single scalar values or vectors. For example, $p = (x, y, z)$ means a space point and $v = (r, g, b)$ may indicate a color image sample.

Second, determine a resolution. Locate the points in grid space, most of the time we deal with a rectangular region. We will get domain cells in a grid space. In this step, we need to find the closest pair of sample points, while not including very many samples in a cell. However, it is fine in some cases to use average values if the measuring area is very wide and two sample points are quite close.

Third, check the sample values to see if a Lipschitz function can be used. If the Lipschitz constant L is too big, we may need to use the local Lipschitz condition by determining the values in range. This process can be done automatically or alternatively. We then do function extension following the theorems and algorithms presented in Chap. 7. We obtain gradually varied or near gradually varied (continuous) functions.

Fourth, use the finite difference method to calculate partial derivatives, then obtain the smoothed function as described in Chap. 7.

Lastly, test the result to see if we need to change the resolution of the grid points. Test to see whether a multilevel or multi resolution method needs to be used.

As we know, the real data application process is much different from the theoretical analysis. An existing mathematical result may not easily produce good, real world results. Different data sets may require some preprocesses. In this chapter, we go into as much detail as possible to explain implementations of the whole algorithm.

8.3 Spatial Data Structures for Data Reconstruction

In computer science, data structures are used to hold the input, or intermediate data for processes. Data structures are usually highly related to the efficiency of algorithms. Even though in mathematics or numerical analysis, researchers focus on algorithms, for real world applications, data structures are extremely important in carrying out the final implementation of programs. In this section, we would like to introduce two of the most common data structures for geometric data processing.

Why are data structures so important to computing? Computing requires computer programs in order to be realized. Data structures lay the foundation for programmers to write programming code, C++ or Java. Different problems will require different data structures (A simple calculation does not usually need a data structure.).

Computer scientist N. Wirth wrote a famous book in the 1970s called *Algorithms +DataStructures = Programs* [25]. The following is what Wirth discusses: To translate an algorithm to a program code is not a simple task. However, when a data structure is selected for the program, the coding becomes straightforward. Notice that this statement was true for structural programming in the 1970s–1990s. Today, object-oriented programming might require modifications in this famous statement. Nevertheless, data structures are still a dominant factor in programming, especially for complex problems such as geometric data processing.

Two data structures are introduced here: (a) the linked-list data structures for graphs, and (b) the mesh or cell data structures for manifold objects. We also assume prior knowledge of the array data structure in multi-dimensions.

The data structures we have covered here will be able to solve the following problems:

1. The domain is rectangular and basic elements of the domain (called cells) are also rectangular in 2D.
2. The domain is in 2D, the cells are triangles or other simplicial shapes (convex polygons).
3. The domain is a 2D manifold in 3D. The cells are triangles or other convex polygons in 3D (not 2D).
4. For digital geometric objects, the domain is a connected component in 3D. The cells are points, line-segment, squares, or cubes (See Fig. 2.5).

8.3.1 Linked-Lists and Linked Graph Data Structures

A linked-list is a collection of nodes that consists of node content and a pointer to the next node. Linked lists are commonly used in computer science (Fig. 8.1).

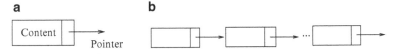

Fig. 8.1 A linked list: (**a**) A node configuration. (**b**) An example of a linked list

A linked list can be very useful in implementing a special data structure, such as the most commonly used *queues* and *stacks*. A *queue* represents a first-in-first-out order of data items, and a *stack* represents a first-in-last-out order.

A graph $G = (V, E)$ can be represented as a set of linked-lists in computers though a so called adjacency list. The adjacency list of a graph, as described in Chap. 2, consists of a collection of linked lists. Each list begins with a node (vertex) followed by all of the adjacent vertices. For the graph in Fig. 8.2, its adjacency list is

$A \rightarrow B \rightarrow C \rightarrow E$
$B \rightarrow A \rightarrow D$
$C \rightarrow A \rightarrow E$
$D \rightarrow B$
$E \rightarrow A \rightarrow C$

The advantage of the adjacency list over the adjacency matrix is that when a graph contains fewer edges, this structure will save a lot of memory space, especially for triangulated meshes where each node only has three adjacent nodes. In addition, the adjacency list uses linked list data structures, which will bring a great deal of

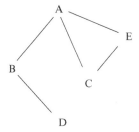

Fig. 8.2 A graph with five vertex

flexibility to triangle deletion (subtraction) and refinement (addition). The adjacency list is for flexible data and complex structures.

8.3.2 Data Structures for Geometric Cells

This data structure first defines points by a vector or a 1D array. For example, points in 3D are (x, y, z). We can also use $p = \{p[0], p[1], p[2]\}$ to represent a point. In C++ or Java,

class point {
int p[3];
}

is used as a "data type" that is able to declare a point in 3D. A face-cell (two-cell), such as a triangle or a rectangle, can be represented by three or four corner points. A solid cell will be a set of face-units.

The typical structure of a data file that holds a solid object will be the following:

$n_0 \quad n_1 \quad n_2$
List of points
List of faces (represented by the number of points and the index of points)
List of 3D-cells (represented by the number of faces and the index of points)

where $n_0 =$ is the number of points, $n_1 =$ is the number of faces, and $n_2 =$ is the number of 3D cells. Here is the actual example in the data file,

5 3 0
0.185223 0.0810056 0.767162
0.163544 0.0766346 0.772773
0.132225 0.0702602 0.766102
0.0778127 0.0591506 0.745873
0.0354809 0.0504085 0.705737
3 1 2 3
3 3 4 0
4 2 3 0 4

This data file has five points, three faces, and zero 3D cells. The five points are listed. Row "3 1 2 3" means that the face is made by three points and the indices are 1, 2, and 3. The three points are

(0.163544, 0.0766346, 0.772773),
(0.132225, 0.0702602, 0.766102),
(0.0778127, 0.0591506, 0.745873).

8.3.3 Linked Cells Data Structures

Computer graphics usually perform individual cell calculations [23]. Each data cell might not have a special requirement in terms of calculations.

In digital geometry, data may need to be traversed from one cell to a neighboring cell. We often use a linked face data list [7] see Fig. 8.3. This is a special digital data structure. The memory space is most effective since the faces are not required to be specifically defined; this is by default in cubic space.

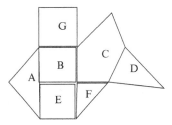

Fig. 8.3 An example of convex-cell decomposition

The linked-face data list:

$A \rightarrow B \rightarrow E$
$B \rightarrow C \rightarrow G \rightarrow E \rightarrow A$
$C \rightarrow D \rightarrow B \rightarrow F$
$D \rightarrow C$
$E \rightarrow A \rightarrow B \rightarrow F$
$F \rightarrow E \rightarrow C$
$G \rightarrow B$

8.4 Implementation of Algorithms

We have discussed the core algorithm in Chap. 7. In this section, we will present detailed data structures and detailed algorithm implementation.

For instance, the program must first store the sample data points, then determine an order of unknown point calculations since different orders may generate different results. The best way is to select the data points in order by distance to sample points. For a point, we mark the order based on the following logic: the shorter the distance, the quicker it is to be selected.

8.4.1 α-Set (Shape) Calculation and Computing Order

We begin with a simple example: A 2D domain with D is $N \times M$ grids (cells). We have a set of sample points that are defined, $f_J : J \rightarrow \{A_1, A2, \cdots, A_m\}$. Even though Chap. 7 proved the algorithm for reconstruction, for real data processing, we can get the order by calculating the shortest distance to the sample set J:

$$ord(x, J) = \min\{d(x, a) | a \in J\}$$

The time complexity of this calculation is $O(|J||D|)$ when $d(x, y)$ is known. In practice, we may only know the index of each point in J. The calculation algorithm can be designed as a simple and fast procedure: Mark all points in J to be "0." Each time, increase the distance using 1 as the marker, i.e. mark the adjacent point of the most recently marked points if it is not marked already. We save the marked points in a queue. Therefore, this algorithm will be linear. The set in computational geometry is called α set [16].

We can see the order form a Voronoi diagram for digital domains. This is also the fastest algorithm for digital α set calculation.

Algorithm 8.1. The following algorithm is for digital Voronoi diagrams with an order selected by the calculations in reconstruction.

Step 1: Mark all points in J to be 0 and put them in a queue Q.
Step 2: $x \leftarrow RemoveHeade(Q)$; put all unmarked adjacent points of x in *Insert Tailer*(Q). Mark these points as the value plus 1 as it is on x.
Step 3: Continue until Q is empty.

Lemma 8.1. *The time complexity of Algorithm 8.1 is $O(n)$.*

If the domain is not rectangular but it is still a mesh, then we can represent each unit as a cell to get a graph and the above algorithm will still be applicable. We discuss this case in following sections and in Chap. 9.

Since the reconstruction of gradually varied functions is not unique, sometimes a random algorithm will generate a more expected result. The two types of randomization are selection of points to be fitted and selection of the value if there is more than one choice. We will discuss the random algorithm for gradually varied fittings in Chap. 10.

8.5 Manifolds and Their Discrete and Digital Forms

A manifold is a topological space. It usually has a measure for defining the distance from two points in the space. We can try to understand a manifold through the following way: a manifold is a connected object where each point has a neighborhood that is similar to a part of Euclidean space (an open set). We can also say that the neighborhood is deformable to an open set in Euclidean space.

For a 2D shape, one could partition the manifold into triangles, rectangles, or polygons. Each of the small shaped-components (two-cell) can be viewed as an open set in 2D without considering its boundaries.

Even through there are many definitions, the best way to describe a manifold in computers is to represent it by a combination of decompositions: triangulations or cubics. The general terminology for this is called cellular complexes or simplicial complexes.

In this section, we mainly discuss the triangulations and cubic representations of manifolds, meaning computer representations of digital and discrete manifolds. We first give the definitions of digital and discrete manifolds. Then, we give the data structure of these manifolds.

8.5.1 Meshes and Discretization of Continuous Manifolds*

Triangulation is often used because triangles are the simplest form to represent an object with an area. All other shapes can be represented by triangular composition, called piecewise linear decomposition [24]. A general format of the local shape of the triangulation decomposition is shown in Fig. 2.3b.

Two drawbacks of triangulations are: (1) The number of cells are much more than general polygons, so the calculation time may be costly, and (2) The sampling error may be large in practical applications (angles in 3D always contain error).

For a 3D object, how do we obtain a triangulated boundary? This process is called meshing in real data collection. A popular technology is called marching cubes [18].

8.5.1.1 Marching Cubes

Marching cubes is the most commonly used method of obtaining a triangulated mesh from a solid object. The algorithm is very simple. To check the type of local configuration of a boundary point, it must be one of the 15 cases in Fig. 8.4a. Then, it just stores the triangles into the file.

3D industrial data such as 3D CT or MRI images are stored as data points. The data points are the set of points $p = (x, y, z)$, which are in Euclidean space. It is easily

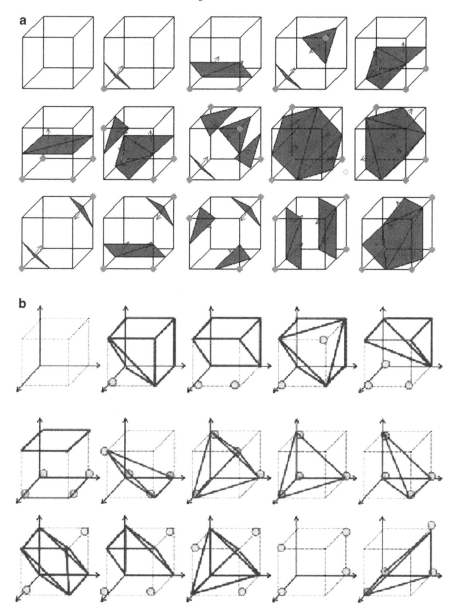

Fig. 8.4 Triangulation from cubes: (**a**) Marching cube, (**b**) Convex-hull boundary

seen as a digital object, so we can see that the industrial data is in a cubic format, i.e. 3D grid space.

The marching cubes algorithm obtains a triangulation for the boundary of the data set by considering the neighborhood of boundary points. See Fig. 8.4a.

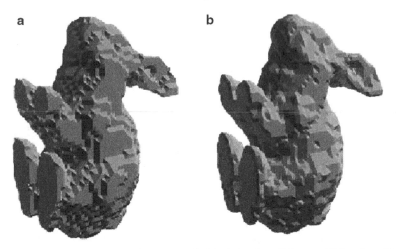

Fig. 8.5 Example of triangulation from cubes (Original data from Princeton 3D benchmark): (**a**) Convex-hull boundary method, (**b**) Marching cube method

8.5.1.2 Convex Hull Representation of 3D Boundaries

Another method we developed uses the convex hull in a cubic cell to represent the boundary of an object. The convex hull formed by certain vertices in a unit cube of digital space is unique. The faces of the convex hull will be cancelled in the inner cubes but will remain on the boundary surface of the 3D solid (object). Unlike the marching cubes method that uses half the points to form meshes (simplicial decomposition), our method preserves the original cube information. For a higher dimensional object, an m-dimensional object in m-dimensional digital space, the boundary of the object in each m-cube is unique. Therefore, the convex hull of the boundary points in an m-cube is also unique. The $(m-1)$-face of the convex hull will represent the boundary of the original object.

In Fig. 8.4b, convex-hull boundary, if there are more than four data points, we will use its complement (blue) configuration.

The convex-hull boundary configuration is a mathematical solution. On the other hand, the marching cube method is a technical solution. We can see that the marching cube method would provide a faulty result in some local points. However, in general, the marching cube method gives more details. See Fig. 8.5.

8.5.2 Digital Manifolds and Its Data Forms*

The advantage of digital manifolds is to directly use a digital object without going through the triangulation meshes. A 2D digital manifold or digital surface can be

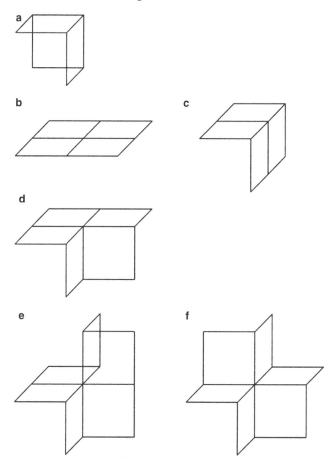

Fig. 8.6 Six types of local digital surface points

viewed as the composition of six local shapes in Fig. 8.6. The digital cubes do not contain calculation errors.

8.5.2.1 Definition of Digital Manifolds

In Chap. 2, we have defined digital space as a set of integer points in the concept of abstract topology. In [23], digital connectivity is defined as the relationship among elements in digital space. For example, four-connectivity and eight-connectivity in 2D. A digital space and its (digital-)connectivity completely determines a digital topology [7].

A similar practice can be traced back to Alexandrov and Hopf in 1935 who first introduced grid cell topology using cellular open sets to define the topology for a

2D grid space (see Chap. 2). Here, we only discuss a simple definition for digital manifolds.

Let us start with digital surfaces. A digital surface in 3D can be viewed as a combination of 2D-cells (one-cells, surface-cells, or surface-unit-cells), which is a $[0,1] \times [0,1]$ square. Based on the theorem of digital surface point classification proved in [7, 12, 13], a digital surface is made by the six (local) shapes shown in Fig. 8.6.

Observe the configuration of these six shapes, we can see that a digital surface has the following properties: (1) every 1D cell (line-cell) is contained by two surface cells (if it is not on the boundary), (2) there is no three-cell, and (3) all one-cells are connected through two-cells.

In general, a digital manifold is a special kind of combinatorial manifold. We can define a digital manifold based on the so called "parallel-moves."

The *parallel-move* is a technique used to extend an i-cell to an $(i+1)$-cell: A and B are two i-cells and A is a parallel-move of B if $\{A, B\}$ is an $(i+1)$-cell. Therefore, k-cells can be defined recursively [7, 11].

Definition 8.1. A connected set of grid points M can be viewed as a digital k-manifold if: (1) any two k-cells are $(k-1)$-connected, (2) every $(k-1)$-cell has only one or two parallel-moves, and (3) M does not contain any $(k+1)$-cells.

8.5.2.2 Advantages of Using Digital Manifolds

Directly using the digital geometry method has considerable advantages since there is no need to premesh the data set into meshes. A theorem by Chen and Rong showed another type of advantage to even differential geometry.

In differential geometry, the Gauss-Bonnet theorem or Gauss-Bonnet formula is one of the most important theorems. It states: Let M be a closed compact two-dimensional (Riemannian) manifold. Then

$$\int_M K dA = 2\pi \chi(M), \tag{8.1}$$

where K is the Gaussian curvature and $\chi(M) = 2 - 2g$, g is a genus, and χ is the Euler characteristic of M.

Use the Gaussian curvature (8.1) and the local configuration of surface points (Fig. 8.6), Chen and Rong proved a digital form of the Gauss-Bonnet theorem presented in Chap. 5 (Sect. 5.4.3): Let M be a closed digital 2D manifold in direct adjacency. The formula for genus is

$$g = 1 + (M_5 + 2M_6 - M_3)/8,$$

where M_i indicates the set of surface-points each of which has i adjacent points on the surface [7, 12, 17].

8.5.3 Two Discretization Methods of 2D

How do we decompose a continuous 2D domain into meshes? For a manifold, there are two types of meshes: (1) Domain meshes, a good example is the Voronoi diagram, and (2) Guiding point meshes, such as Delaunay triangulations.

A manifold can be defined by CW-complexes in topology. We could assume that a fine decomposition of simplexes can represent the manifold, i.e. the linear decomposition (piecewise linear representation) [24].

Since D may only contain few points, we cannot assume that there is linearity among the guiding points, regardless of whether we could assume linearity among domain points. If we have to use the triangulations in guiding points, there is a reasonable way to do this. We could find all the possible simplex decompositions and then find the average to approximate the original. However, it would cost exponential time to accomplish this task because of the number of different forms of simplex decompositions for n points.

In terms of sample points issue to the domain decomposition, domain meshes use the sample point locations as the center points, the value at the center point will represent the cell's value. For example, each point in the Voronoi polygon will have the same value. On the other hand, point meshes will use the vertex values (the sample values) to interpolate the value in side of a triangular cell. The two methods are linear interpolations.

A way that may overcome or solve the problem is to find a nonlinear reconstruction method. We discussed this case in Chap. 7 where we focus on the digital-discrete method. More discussion about the nonlinear method will be presented in Sect. 8.8.2.

In addition, gradually varied functions can also make interpolations based on both the domain meshes and the point meshes since these two meshes are graphs.

8.6 Function Reconstruction on Manifolds

Functions on manifolds relate to the study of functions on a domain that is not a flat rectangle. For example, making colors on a sphere or painting a car. Gradually varied surface reconstruction does not rely on the shape of the domain and it is not restricted by simplicial decomposition. As long as the domain can be described as a graph, our algorithms will apply. However, the actual implementation will be much more difficult. In the above sections, we have discussed two types of algorithms for a rectangular domain. One is the complete gradually varied function (GVF) fitting and the other is the reconstruction of the best fit based on the gradual variation and finite difference methods.

The following is the implementation of the method for digital-discrete surface fittings on manifolds (triangulated representation of the domain).

So far we have dealt with a function or surface on 2D rectangle domains or a two dimensional shape in 3D space. These are not difficult to understand. However, considering a function on a manifold is relatively hard. The best way of giving an easy explanation is painting a car. The surface of a car is a 2D manifold and the new color is the function. We can also view this as surfaces on surfaces.

Gradually varied surface reconstruction does not rely on the shape of a domain and it is not restricted by triangulation or simplicial decomposition. As long as the domain can be described as a graph, this algorithms will apply.

So, if we use gradually varied methods, the key to function reconstruction includes the calculation of the fitting order for unknown points. The algorithm of α-shape for the cells still applies.

Since the gradually varied method is the method for the continuous function on manifold. In Chap. 9, we go one step deeper to get a local harmonic function reconstruction on manifolds.

We use techniques related to the fitting include the vertex fitting or face fitting for 2D manifolds. Vertex fitting puts the values on the vertices; face fitting puts the value (color) on the face (two-cells). We have discussed these two data presentation methods in the above sections.

The differences of these methods in generating different base graphs as data structure are: (1) Using the Voronoi type, each face-cell is a vertex in the graph, and (2) Using Delaunay diagrams, each point-cell is a vertex in the graph.

8.6.1 Data Reconstruction for Triangle Meshes

The following is the implementation of the method for digital-discrete surface fitting on manifolds (triangulated representation of the domain).

We can have four algorithms related to continuous (or smooth) functions on manifolds. This is because we have four cases: (1) The GVF extension on point space, corresponding to Delaunay triangulations; the values are integers (or A_i's). (2) The GVF fitting on point space, the fitted data are real numbers using the Lipschitz constant to determine the values. (3) The GVF extension on face (2D-cell) space, corresponding to Voronoi decomposition; the values are integers. (4) The GVF fitting on face (2D-cell) space, the fitted data are real numbers.

Figure 8.7 shows the original data set, a 3D image. In Fig. 8.8a we show an example using ManifoldIntGVF in a 2D closed manifold (can be viewed as the boundary for a 3D object).

We have modified the data by adding some triangles to make the graph connected. Then, we put four values on the four vertices. Figure 8.8b shows the result. In order to avoid being fooled by the 3D display causing the color to change, we have used two additional sets of guiding points. Each of these uses only three guiding points. The second set of guiding points has a pair of values swapped with the first set (the second set is the same as the first set with a pair of values swapped). The values are posted in the pictures. See Fig. 8.8.

It is very difficult to select values for a relatively large set of sample points for a complete GVF fitting, such as the ManifoldIntGVF, to satisfy the condition of

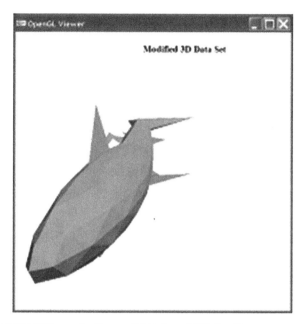

Fig. 8.7 The original 3D image for the manifold integer GVF algorithm

gradual variation (or Lipschitz condition). For ManifoldRealGVF, we can choose any data values. Figure 8.9 shows the results with six guiding points.

Now we obtained smooth functions based on this GVF fitting. We can use the Catmull-Clark method, possibly with another method [5]. A small problem we encountered is the display. In order to make the display for the vertices, we use the average value of three vertices as the display value on each triangle. We will make smaller triangles later on for a more precise display. This problem will disappear in the cell-based GVF. However, the graph needs to be calculated first in order to obtain the cells and their adjacent cells.

8.6.2 Data Reconstruction for Digitally Represented Objects

Triangulation or trianguled meshes are often selected to represent a object. This is because that triangulation is flexible for any complex shapes. Using digital or cubic representation, advantages are that it does not require a mesh. It can be treated directly.

The adjacency list of each cell is also simple. Only six of these exist in 3D space. The data structure will be the same as it is in triangulated objects.

In the above sections, we have discussed two types of algorithms for a 2D domain in a plane. One is the complete gradually varied function (GVF) fitting and the

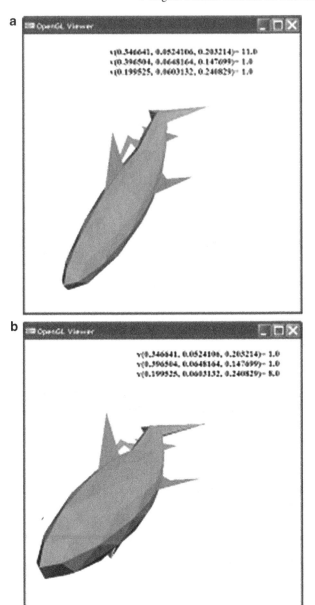

Fig. 8.8 The GVF Algorithm based on three sample point sets: (**a**) The gradual variation from *back* to *front*, (**b**) The gradual variation from the *top* to the *right side*

Fig. 8.9 The ManifoldRealGVF algorithm using six guiding points

other is to reconstruct the best fit based on gradual variation (if the gradual variation condition is not met).

For digital types of fitting algorithms such as using the scale $\{A_1, \cdots, A_m\}$, denoted by ManifoldCellIntGVF, we could use the best fit strategy. For the Lipschitz type, we need to calculate the slope, the algorithm is denoted by ManifoldCellReal-GVF. For simplicity, the maximum slope of two sample points is used to determine the value levels. This treatment will guarantee that the sample point set satisfies the gradual variation condition and the Lipschitz condition (Fig. 8.10).

After the values at all vertices are known for each vertex, we can construct the smooth functions for geodesic paths that cover all edges and link to a vertex. Then, we can use vector space calculus to get the unknown points. A partition of unity may be used here for getting a smooth solution on the partition edges (similar to some finite element algorithms). We can also use the existing methods for obtaining the smooth functions, see Chap. 12.

8.7 Volume Data Reconstruction

The volume reconstruction can be the same as the surface reconstruction since a volume can be represented as a graph. However, when the volume data contains some hard interfaces or fault lines, fitting a smoothed function for the entire volume may "erase" the hard interface.

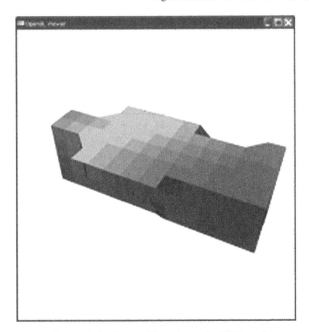

Fig. 8.10 Using seven points to fit the data on 3D surfaces: the GVF result

To perform an appropriate data fitting, the idea is to use some controlling inter-faces. For instance, the separating surface between two layers in geological data. We can first fit the surfaces then use the morphing or lofting technology (also see Chap. 5) to fit the entire volume. See Fig. 8.11.

In [14], we described such a method to reconstruct a velocity volume. In an ap-plication, we plan to reconstruct a $41 \times 20 \times 201$ velocity array. We only have a few data traces to serve as sample points, each trace contains data from top to bottom. We have designed the following procedure.

Procedure of Volume Data Reconstruction:

Step 1:   Velocity layer classification. In this step, we first apply an image segmen-tation (see Chap. 12) method on data traces. We now have five layers.

Step 2:   Boundary curve fitting and interface fitting. The goal of this step is to fit or find every interface. We first fit the boundary of the surface to a rational function, which gives us a best "looking" answer. Once the boundary fitting curve is fitted, we use a gradually varied surface fitting algorithm to obtain the layer interfaces.

Step 3:   Surface Lofting. A lofted surface is obtained by the interpolation of two interface surfaces. In this application, we arrived at 171 lofted surfaces. Each lofted surface was treated as a small layer.

Step 4:   Gradually varied fill in lofted surfaces. A lofted surface only gives the ge-ometric locations in which each point should be assigned a "continuous" velocity value.

Step 5:   Since some points were not in the lofted surfaces, we simply assigned a
value to these points using 1-D linear interpolation.

8.8 Practical Problems and Solving Methodology

The clear statements on the philosophy and methodology of digital-discrete methods are important. Two things are basic. The new method has a solid mathematical
foundation and the method is relatively simple. We have presented some advantages
and disadvantages of the digital-discrete method. In this section, we discuss in depth
when the digital-discrete method should be used.

Even though we have explained the reasons for inventing a new method of
smooth function reconstruction in Chap. 7, there is still a need for further discussion.
For instance, one could ask: With a piecewise-linear method such as triangulation,
apply the Weierstrass theorem or Bernstein polynomials; it seems that we can have
a smooth function that infinitively approaches a function that has the same values at
the guiding points. The task could be done in such a way. Why do we need a new
digital-discrete method for such a task?

In this section, we will start to give more detailed reasons. Some of these can be
different view points or issues for discussion. Philosophical thinking and methodology are important to the future development of science and technology.

The most common problem in data reconstruction is fitting a function based
on the observations of some sample (guiding) points. In this section, we provide
a methodological point of view of digital-discrete surface reconstruction. We explain the new method along with why it is a truly universal and nonlinear method,
unlike most popular methods, which are linear and restricted. This chapter focuses
on what the surface reconstruction problem is and why the digital-discrete method
is important and necessary, along with how it can be accomplished.

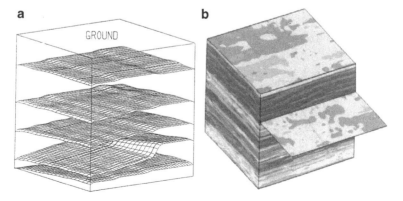

Fig. 8.11 Volume data reconstruction: (**a**) interface interpolation by layers, (**b**) the filled volume

8.8.1 Weakness of Piecewise-Linear Methods

As a classical method, piecewise-linear methods have dominated the practical calculations for more than 100 years.

In general, the classical geometric method uses domain decomposition to assist in
obtaining the numerical interpolation that only contains random sample points. This
is because we need to determine the region that a sample point affects. Therefore,
the Voronoi decomposition and its dual correspondence, the Delaunay triangulation,
are the most popular in practice.

The Voronoi decomposition provides a piece-wise linear approximation to a manifold. This is also regarded as the classical discrete (data) reconstruction method and
can be used with piecewise polynomials.

For example, when we have a Delaunay triangulation based on guiding points,
we could define a spline function since we know all the values of the vertices or we
could add a partial differential equation constraint on the sample points using the
finite elements method [4]. As a result, some researchers would believe scientists
have already found the perfect solution to the problem.

However, this is not the case in the real world. When we use the Voronoi decomposition for sample points, we are simultaneously assuming that the guiding points
are "linearly separable." However, if we know that the unfitted function is not linear,
we can conclude that the guiding points cannot be easily separated linearly or that
we do not know the separation contour curve based on the guiding points if they
are not dense enough. Euclidean space is a Jordan space because it is dense. If the
guiding points are not dense and are randomly arranged, then the Voronoi decomposition or any other triangulation will generate false domain decompositions. This
is the situation in Fig. 8.11.

Even though a piecewise linear method can be applied, there are so many different piecewise cases or triangulations. How do we know which one is right? Even
though the Delaunay method is the most reasonable, what if the sample points are
located on just a line? In other cases, what will be the best selection from the different triangulations? see Fig. 8.12.

When the number of sample points increases, the number of triangulations will
increase dramatically, see Figs. 8.13 and 8.14.

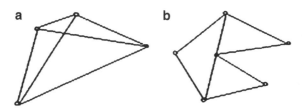

Fig. 8.12 Valid cases but classical methods cannot treat properly. (**a**) A nonlinear (non-Jordan)
case. The point connections are not linearly separable. (**b**) Non-regular or improper mesh case; the
finite elements method does not allow this type of decomposition

Voronoi and Delaunay decomposition may not work the best when sample data points are selected in an elliptical, circular, or triangle boundary.

8.8.2 The Digital-Discrete Method is a Nonlinear Method

In Chap. 3, we talked about how the digital function is a half numerical and half symbolical method. To say this, we recall the nature of digital functions. We can map $\{1, 2, \cdots, m\}$ naturally to $\{A_1, A_2, \cdots, A_m\}$ and the fitting only cares about the symbols of $A_1, A_2, \cdots, A_m$. However, $A_1, A_2, \cdots, A_m$ form an order. This has a numerical meaning. Because of these, digital functions can apply to graphs, and graphs

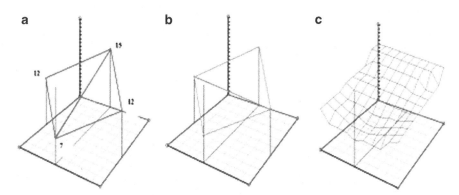

Fig. 8.13 Piecewise linear vs. GVF: (**a**) and (**b**) Two piecewise linear interpolations but we do not know which one is correct. (**c**) The GVF interpolation results show a reasonable non-linear fitting

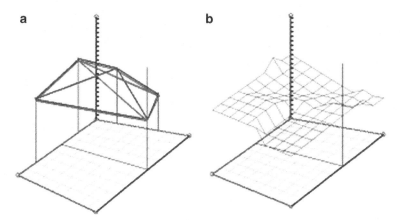

Fig. 8.14 A case with five sample points will result in ten piece-wise linear interpolations: (**a**) Two interpolations are shown, (**b**) A GVF interpolation

are usually a nonlinear space. It is different from Euclidean space, which is a linear space in a continuous sense. See Figs. 8.13(c) and 8.14(b).

On the other hand, the digital only method uses only $\{1, 2, \cdots, m\}$ and has some problems in representing derivatives. It will not be able to represent the real world values since the largest derivative value will be 1. See Chaps. 3 and 7.

Therefore, gradually varied functions can deal with the general sense of derivatives, a discrete method on graphs. In addition, the actual value and location can be restored to each site to make exact interpolations as needed.

Data reconstruction is used to fit a function based on the observations of some sample (guiding) points. In Chap. 7, we present some applications of using recently developed algorithms for smooth-continuous data reconstruction based on the digital-discrete method [6, 8, 9]. The classical discrete method for data reconstruction is based on domain decomposition according to guiding (or sample) points. Then, the spline method (for polynomials) is used to fit the data, see Chap. 6.

Some successful methods have been discovered or proposed to solve the problem including the Voronoi-based surface method and the moving least squares method [1–3, 12, 17, 20, 21, 29]. Even the finite elements method (for PDE) can be used in data reconstruction, see Chap. 11. However, all of the existing methods have had an assumption that is the domain is linear. Even though Euclidean space is linear, the points from random sampling might not be linear since the connectivity among these sample points might not be linear.

The new method is based on the gradually varied function that does not assume the property of being linearly separable among guiding points, i.e. no domain decomposition methods are needed. We also demonstrate the flexibility of the new method and its potential to solve a variety of problems. In Chaps. 9 and 10, the examples include some real data from water well logs and harmonic functions on closed 2D manifolds. This method can be easily extended to higher multi-dimensions.

The digital-discrete method can deal with smooth functions on both the piecewise linear manifold and a non-Jordon graph. The non-Jordon graphs will only affect the calculations in terms of speed. We discovered that the major difference between our methods and existing methods is that the former is a true nonlinear approach [9].

8.8.3 Digital-Discrete Method as a Universal Method

The digital-discrete method can be applied to many applications without limitation. Even though it is new and has many aspects that require further investigation, it is very general just like the spline method. On the other hand, the Lipschitz extension has some simplicity in theory. However, when we apply it in real data processes, it may not produce a good result (see last section of Chap. 7). This is because every function in a bounded area is Lipschitz, but the Lipschitz-constant could be very large and we could not directly use it.

A local Lipschitz function on a manifold might not be able to be fully implemented by one set of rational or real numbers $A_1 < A_2 < \cdots < A_n$. A common method in image processing is called pyramid representation based on the intensity of an image. For a time sequence it is also well-known that a curve can be represented by a tree. It may be related to the Morse (height) function in topology. Using a tree instead of a chain for range space would solve the problem completely in theory. This method was very well prepared in 1990, when Chen proved a theorem that states that any graph can be normally immersed into a tree, see the last section of Chap. 4 [7]. This means that there is a gradually varied extension from a graph to a tree. A more detailed description of this approach can be found in Chaps. 4 and 5. Two values A_i and A_j that have the same value but are located in different braches of a tree represent different elements. They are not the same.

Smooth gradually varied functions can also work with other numerical methods. It adds more flexibility to the new method, see Chap. 11.

Assume a domain is linearly separable, such as the Jordan domain. If we need a local region, we can first fit the boundary of the region and then fit the inner points. Our discrete algorithm does not limit the order of the reconstruction. So the gradually varied function fitting is both a local and global method. In the above sections, we show several examples using GVF and we have conducted more real data processing in Chaps. 9 and 10.

8.9 Remarks

We have presented several algorithms and examples using the new digital-discrete method in this chapter. The solution of the gradually varied function in discrete space is for more general problems without Lipschitz-condition limitation. This method is practical and is different from the research by Fifeman, Whitney and McShane [14]. The gradually varied function is highly related to the Lipschitz function and the local Lipschitz function. To get a smoothed function using gradual variation is a long time goal of our research. Some theoretical attempts have been made before, but struggled in the actual implementation [10].

Many thanks to Dr. Diane Lingrand at Polytech Sophia for providing the marching cubes image. Some 3D examples are from Princeton 3D Benchmark. Figure 8.11 was made in 1990 by L. Chen and L.C. Zheng *et al* in a project in Nanjing, China.

References

1. Amenta N, Kil YJ (2004) Defining point-set surfaces. In: ACM SIGGRAPH 2004 Papers, Los Angeles, Aug 08–12 2004
2. Amenta N, Bern M, Kamvysselis M (1998) A new Voronoi-based surface reconstruction algorithm. In: Proceedings of SIGGRAPH '98, San Francisco, July 1998, pp 415–422
3. Belytschko T, Krongauz Y, Organ D, Fleming M, Krysl P (1996) Meshless methods: an overview and recent developments. Comput Methods Appl Mech Eng 139(1–4):3–47
4. Brenner SC, Scott LR (1994) The mathematical theory of finite element methods. Springer, New York

5. Catmull E, Clark J (1978) Recursively generated B-spline surfaces on arbitrary topological meshes. Comput Aided Des 10(6):350–355
6. Chen L (1990) The necessary and sufficient condition and the efficient algorithms for gradually varied fill. Chin Sci Bull 35(10):870–873
7. Chen L (2004) Discrete surfaces and manifolds. Scientific and Practical Computing, Rockville
8. Chen L (2010) A digital-discrete method for smooth-continuous data reconstruction. Capital Science 2010 of The Washington Academy of Sciences and Its Affiliates, Washington
9. Chen L (2010) Applications of the digital-discrete method in smooth-continuous data reconstruction. http://arxiv.org/ftp/arxiv/papers/1002/1002.2367.pdf
10. Chen L, Adjei O (2004) Lambda-connected segmentation and fitting. Proc IEEE Int Conf Syst Man Cybern 4:3500–3506
11. Chen L, Zhang J (1993) Digital manifolds: a intuitive definition and some properties. In: The proceedings of the second ACM/SIGGRAPH symposium on solid modeling and applications, Montreal, pp 459–460
12. Chen L, Zhang J (1993) Classifcation of simple surface points and a global theorem for simple closed surfaces in three dimensional digital spaces. Proceedings of the SPIE's Conference on Vision Geometry. Boston, Massachusetts, USA, SPIE 2060:179–188
13. Chen L, Cooley H, Zhang J (1999) The equivalence between two definitions of digital surfaces. Inf Sci 115:201–220
14. Chen L, Cooley DH, Zhang L (1998) An intelligent data fitting technique for 3D velocity reconstruction, application and science of computational intelligence. Proc SPIE 3390:103–112
15. Chen L, Rong Y (2010) Digital topological method for computing genus and the Betti numbers. Topol Appl 157(12):1931–1936
16. Edelsbrunner H, Kirkpatrick D, Seidel R (1983) On the shape of a set of points in the plane. IEEE Trans Inf Theory 29(4):551–559
17. Fefferman C (2009) Whitney's extension problems and interpolation of data. Bull Am Math Soc 46:207–220
18. Lorensen WE, Cline HE (1987) Marching cubes: a high resolution 3D surface construction algorithm. Comput Graph 21(4):163–169
19. Lancaster P, Salkauskas K (1981) Surfaces generated by moving least squares methods. Math Comput 87:141–158
20. Levin D (2003) Mesh-independent surface interpolation. In: Brunnett G, Hamann B, Mueller K, Linsen L (eds) Geometric modeling for scientific visualization. Springer, New York
21. McLain DH (1976) Two dimensional interpolation from random data. Comput J 19:178–181
22. Mount DM (2002) Computational geometry, UMD lecture notes CMSC 754. http://www.cs.umd.edu/~mount/754/Lects/754lects.pdf
23. Rosenfeld A, Kak AC (1982) Digital picture processing. Academic, New York
24. Thurston W (1997) Three-dimensional geometry and topology. Princeton University Press, Princeton
25. Wirth N (1970) Algorithms + data structures = programs. Prentice-Hall, Englewood Cliffs

Chapter 9
Harmonic Functions for Data Reconstruction on 3D Manifolds

Abstract The goal of smooth function reconstruction on a 2D or 3D manifold is to obtain a smooth function on surfaces or higher dimensional manifolds. It is a common problem in computer graphics and computational mathematics, especially in civil engineering including structural analysis of solid objects. In this chapter, we introduce a new method using harmonic functions for solving this problem. This method contains the following steps: (1) Partition the boundary surfaces of the 3D manifold based on sample points so that each sample point is on the edge of the partition. (2) Use gradually varied interpolation on the edges so that each point on the edge will be assigned a value. In addition, all values on the edge are gradually varied. (3) Use discrete harmonic functions to fit the unknown points, i.e. the points inside each partition patch. This solution of the fitting becomes the piecewise harmonic function.

9.1 Overview: Real Data on Manifolds

Obtaining a smooth function on a 2D or 3D manifold is a common problem in computer graphics and computational mathematics. In computer graphics, smooth data reconstruction on 2D or 3D manifolds usually refers to subdivision problems [3]. Such a method is only valid when based on designed or desired shapes (what types of shapes we wanted is known) or dense sample points. The manifold usually needs to be triangulated into meshes (or patches) where each node on the mesh has an initial value. While the mesh is refined, the algorithm provides a smooth function on the redefined manifolds.

Most of real data reconstruction is under the condition that we do not know what the shape should be. In addition, when data points are not dense and the original mesh is not allowed to be changed, how do we make a "continuous and/or smooth" reconstruction? [4]

In Chap. 8, we used gradually varied functions to fit a closed 2D surface. In this chapter, we present a new method using harmonic functions to solve the problem.

L.M. Chen, *Digital Functions and Data Reconstruction: Digital-Discrete Methods*,   123
DOI 10.1007/978-1-4614-5638-4_9, © Springer Science+Business Media, LLC 2013

This chapter will mainly focus on surface reconstruction. We also present a method for data reconstruction on solid objects in 3D.

The fitted function will be a harmonic or local harmonic function in each partitioned area. The function on edge will be "near" continuous (or "near" gradually varied). If we need a smooth surface on the manifold, we can apply subdivision algorithms. We discuss this option in Chap. 12.

The new method we present may eliminate some of the uses of triangulation. In computer science, we have relied on this foundation for at least 30 years. In the past, people usually used triangulation for data reconstruction. Here, we employ harmonic functions, a generalization of triangulation, because linear is a form of harmonic. Therefore, local harmonic initialization is more sophisticated than triangulation.

Our method for a 3D function extension contains the following steps:

(1) Partition the boundary surfaces of the 3D manifold based on sample points so that each sample point is on the edge of the partition.
(2) Use gradually varied interpolation on the edges so that each point on the edge will be assigned a value. In addition, all values on the edge are gradually varied [5, 6].
(3) Use discrete harmonic functions to fit the unknown points, i.e. the points inside each partition patch. Finally, we can use the boundary surface to do a harmonic reconstruction for the original 3D manifold.

This result of the fitting becomes the piecewise harmonic function on the manifold.

9.2 Harmonic Functions and Discrete Harmonic Functions

In this section, we briefly review harmonic functions and discrete harmonic functions. Harmonic functions have important applications in science and engineering. The solution to the Dirichlet problem is harmonic. In addition, the linear function is harmonic. As an example, the function for electric potential due to a line charge

$$f(x_1, x_2) = \ln(x^2 + y^2).$$

is harmonic.

Harmonic functions also play an important role in solving the famous minimal surfaces problem [10].

9.2.1 Harmonic Functions

A function $f(x, y)$ is said to harmonic if

$$\frac{\partial^2 f}{\partial x^2} + \frac{\partial^2 f}{\partial y^2} = 0 \qquad (9.1)$$

Harmonic functions have some important properties: (1) the maximum and minimum values must be on the boundary of the domain (if f is not a constant). (2) The average of the values in a neighborhood circle equals the value at the center of the circle.

Harmonic functions are also fundamental in mathematics because they relate to the Dirichlet problem: given a continuous function f on boundary ∂D of D, is there a differentiable extension F of f on D? The solution to this problem is unique.

The method used in finding the solution is called the variational principle, which laid out the foundation for functional analysis. The Dirichlet problem is also important in mathematical physics. We can see that gradually varied fitting is similar to the Dirichlet problem, but the solution for gradually varied functions might not exist even if the boundary is gradually varied. We will present a more detailed discussion in Chap. 11 [8].

For a simple region D and its boundary J, we have the following [12, 13]:

Theorem 9.1. *For a bounded region D and its boundary J, if f on J is continuous, then there is a unique harmonic extension F of f such that the extension is harmonic in $D - J$.*

More generally, the solution for the Dirichlet problem is [12]:

Theorem 9.2. *Let us take a 2D piecewise linear manifold with multiple closed simple curves as its boundaries (the genus of the manifold could be $g > 0$). If the function f on each of these curves is continuous, then there will be a harmonic extension of the function f on the whole manifold.*

9.2.2 Discrete Harmonic Functions

Let $f : D \to R$ be a function on D where $G = (D, E)$ is a graph. f is said to be discretely harmonic if $f(x)$ is equal to the average value for all adjacent vertices of x [2, 14]. In other words, if f_i is the value on vertex i, then

$$a_{1,i} f_1 + \cdots + a_{i-1,i} f_{i-1} + a_{i+1,i} f_{i+1} + \cdots + a_{n,i} f_n = \left(\sum_{j \neq i} a_{i,j} \right) f_i, \qquad (9.2)$$

where $A_{n \times n} = \{a_{i,j}\}$ is the adjacency matrix meaning that $a_{i,j} = 1$ if vertex i is adjacent to vertex j, otherwise $a_{i,j} = 0$.

To actually build the matrix of the system of linear equations, we can assume $|D| = n$, $|J| = k$, and $m = n - k$. Without lossing generality, we can let $J = \{m + 1, \cdots, n\}$. So $f_{m+1}, \cdots, f_n$ are the sample points. Their values are known. The system of linear equations will be

$$\begin{pmatrix} \sum_{j\neq 1}a_{1,j} & -a_{1,2} & \cdots & -a_{1,m} \\ -a_{1,j} & \sum_{j\neq 2}a_{2,j} & \cdots & -a_{2,m} \\ \cdots & \cdots & \cdots\cdots & \\ -a_{m,1} & -a_{m,2} & \cdots & \sum_{j\neq m}a_{m,j} \end{pmatrix} \begin{pmatrix} f_1 \\ f_2 \\ \cdots \\ f_m \end{pmatrix} = \begin{pmatrix} \sum_{j=m+1,n}a_{1,j} \\ \sum_{j=m+1,n}a_{2,j} \\ \cdots \\ \sum_{j=m+1,n}a_{m,j} \end{pmatrix} \qquad (9.3)$$

Based on this definition, discrete harmonic extension can be solved by the system of linear equations. See the following example Fig. 9.1a where the value at a vertex is putted into the node.

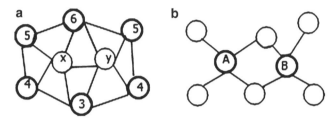

Fig. 9.1 Discrete harmonic extension: (**a**) Guiding points at the edge, (**b**) Guiding points at the center

For each unknown point, we can establish an equation, so we have two equations for two variables.

$$3+4+5+6+y=5x$$
$$3+4+5+6+x=5y$$

So $y=5x-18$; $3+4+5+6+x=5(5x-18)$; $24x=6\cdot 18$, $x=4.5$; and $y=4.5$. We can also solve the equation indicated in Fig. 9.1b.

This solution is unique. A theorem will be presented in this chapter, which we will explain in the next section. This result matches with the case in continuous space, known as the Dirichlet problem, which also has a unique solution [10, 12].

If we add a sample point x at the center of D, then $f(x)$ may not equal the fitted function without $(x, f(x))$. Therefore, the fitted function might not be harmonic at x. These points are called poles and are not harmonic.

Based on Theorem 9.2, the discrete version of this theorem can be validated by taking a planar graph D and creating a small circular neighborhood $C(x)$ around each pole in J. We have the following: There is an extension of f_J, called f_D, such that f_D is harmonic at all points in $D-J$.

Our purpose for data construction is to get a function that is smooth at each "pole" point in J. This type of construction is against our objective, see Fig. 9.2.

No one wants this type as a fitted surface except when dealing with a tent or airport custom podium. Sometimes, mathematical validity and beauty do not fit a problem's purpose. Therefore, harmonic functions must be used in other appropriate ways.

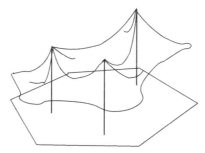

Fig. 9.2 Direct harmonic reconstruction on randomly arranged points only generates a tent type of surface (Assume values on boundary samples are low)

The new method designed in this chapter begins with the gradually varied function method and then uses harmonic reconstruction. We partition a manifold into polygons where each sample point is located on the edge of the polygon. A gradually varied fitting is then performed on the edge of the polygon. Then, we apply the discrete harmonic interpolation.

9.2.3 Principle of Finite Difference Based on Discretization and Extension*

A general format for the harmonic equation is called Poisson's equation

$$\frac{\partial^2 f}{\partial x^2} + \frac{\partial^2 f}{\partial y^2} = g(x, y). \tag{9.4}$$

This equation is essential to both elliptical and parabolic partial differential equations. We discuss these in Chap. 10. The finite difference method for this equation will provide a solid foundation to our numerical solutions for harmonic extension.

In this subsection, we present the theorem for unique extensions. This theorem is applicable in many difference methods for connected domains since it requires the use of a linear system of equations to solve the interpolation problem.

Let us first introduce a concept called the diagonally dominant matrix in linear algebra.

Definition 9.1. A matrix $A[n \times n]$ is said to be diagonally dominant if for each row we have

$$|a_{ii}| \geq \sum_{j \neq i} |a_{ij}|; \, i, j = 1, 2, \cdot, n.$$

and

$$|a_{ii}| > \sum_{j \neq i} |a_{ij}|; \text{ for at least one } i$$

Based on Sect. 2.2.2, the finite difference equation derived from (9.2) is (Δ_x $=\Delta_y=1$)

$$f(x-1,y)+f(x+1,y)+f(x,y-1)+f(x,y+1)-4f(x,y)=g(x,y) \qquad (9.5)$$

If point (x,y) is adjacent to a point in the sample point set J, for example $(x-1,y)$ is on the boundary of D, then $f(x-1,y)=c$ is known. We have

$$f(x+1,y)+f(x,y-1)+f(x,y+1)-4f(x,y)=g(x,y)-c \qquad (9.6)$$

This means that if $J \neq \emptyset$, then the absolute value of the coefficient of $f(x,y)$ is greater than the summation of what remains. Therefore, the finite difference discretization of Poisson's equation is diagonally dominant [11].

We now verify that the matrix is also irreducible. A graph is irreducible if for any two vertices p and q, there exists a path from p to q. Since D is connected, this path always exists. We can see that the graph is irreducible. The adjacency matrix is irreducible if the graph is irreducible. Therefore, the matrix for finite difference discretization is irreducible and diagonally dominant.

Theorem 9.3. *If a matrix $A[n \times n]$ is strictly diagonally dominant or irreducibly diagonally dominant, then it is nonsingular.*

To prove this theorem, we need to use the Gershgorin circle theorem [13, 19, 20]. This theorem guarantees that the discrete harmonic extension will have a unique solution.

Theorem 9.4. *The solution to a harmonic equation of nonempty guiding points is unique.*

The above discussion also relates to the principle of interpolation and extension of the finite difference based method.

9.3 Piecewise Harmonic Functions and Reconstruction

In Sect. 9.2, we explained why harmonic functions are used in data reconstruction when a domain boundary is known. In order to use the discrete harmonic function in data reconstruction, we must first determine the values on a boundary curve. According to the solution of the Dirichlet problem, the values on the boundary are continuous. For discrete cases, we use gradually varied functions to replace continuous functions. We restate the definition of reconstruction here:

Defining Discrete Surface Fitting: Given $J \subseteq D$, and $f : J \to \{A_1, A_2, \ldots A_n\}$, decide if there exists an $F : D \to \{A_1, A_2, \ldots, A_n\}$ such that F is gradually varied where $f(x) = F(x), x \in J$.

We presented a method to obtain the gradually varied functions on manifolds in Chap. 8. In Chap. 7, we implemented the algorithms for smooth data reconstruction in 2D rectangle domains and gradually varied reconstruction for 2D manifolds [13].

To use (discrete) harmonic functions in data reconstruction, we need smoother surfaces on manifolds. Since we usually have randomly sampled points, we need a way of finding a partition for the domain based on the sample points. Delaunay triangulation is one way of doing this. However, the shape of each domain may not necessarily be a triangle. We can partition the domain randomly using an algorithm. We will discuss the details in next subsections [9].

9.3.1 Procedures for One Boundary Curve

Let D be a manifold and D_0 be a component in D. First, we present the algorithm for the harmonic extension of a domain with a simple curve as the boundary of a region D_0, called ∂D_0, on a manifold D. Figure 9.3 shows an example where $J = \partial D_0$.

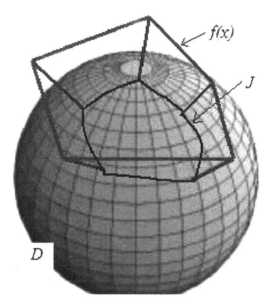

Fig. 9.3 An example of 2D manifold D, a closed curve J, and $f : J \rightarrow \{A_1, A_2, \cdots, A_n\}$

We show an algorithm in Algorithm 9.1. This algorithm is an iterated. Assume the value on the boundary is gradually varied, the iterated procedure will stop when the fitted function reaches the limited error range as given at the beginning of the procedure. This algorithm can also use a system of linear equations to find the solution.

Algorithm 9.1. Input: A closed curve C on manifold M. Let $f_C : C \rightarrow \{A_1, \cdots, A_m\}$ be gradually varied on C. Output: A harmonic extension of M (for two components of $M - C$).

Step 1:   Do a gradually varied interpolation of M.
Step 2:   Use the result as the initial value to do an iteration based on equation (9.2)
Step 3:   Repeat Step 2 until the error is limited as specified.

Using a data set, see Fig. 9.4(a), (b), we apply the Algorithm 9.1 to the data. we have iterated the process 100 times to get the fitting results for both the inside and outside of the closed curve. Comparing gradually varied fitting and harmonic fitting results in Fig. 9.5(a), (b), we can see that the harmonic fitting gives more detailed values for cells. In other words, the harmonic fitting is smoother than the gradually varied fitting. This is in accordance with the theory: Gradually varied surfaces are continuous and harmonic surfaces are smooth.

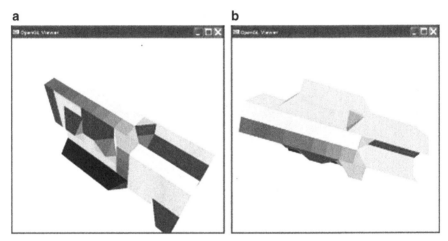

a   **b**

Fig. 9.4 The selected cells form a boundary curve that is gradually varied: (**a**) and (**b**) are two displays for the guiding points (cells)

9.3.2 Procedures for Randomly Sampled Points

It is difficult to make a smooth function on a closed manifold if manifold refinement is not allowed. Some methods invented in the past are based on relatively dense sample points [1, 17, 18]. GVF is appropriate for the continuous function; it is difficult to apply a Taylor expansion locally with an arbitrary neighborhood without a dense domain. Local Euclidean blocks are usually adopted by computer graphics in subdivision problems [3] .

The algorithm here partitions the manifold into several components so that each sample point is at the edge (boundary) of the partition. The edge of the component is a simple closed curve (or path in terms of graphs). Each closed curve contains original sample points. We also assume that the component is simply connected.

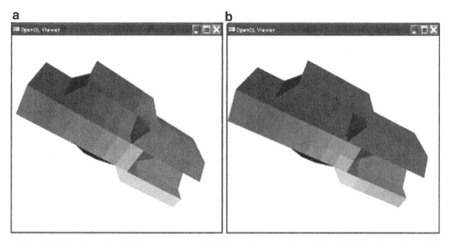

Fig. 9.5 Gradually varied fitting vs. harmonic fitting: (**a**) The Harmonic fitting based on GVF, (**b**) The direct GVF result

Then, we find the gradually varied interpolation on these paths/curves (each curve one-by-one). Therefore, we will have data on all boundaries of all components; the fitted data points are all gradually varied. The inside of each component does not contain any data or sample points except boundary edges. Finally, we use the harmonic function to fit each component using a simple iterating process so we will have a locally harmonic function on the surface. If we need a smooth function on the surface, we would need to then apply a subdivision algorithm to the above solution [9].

We first present this algorithm in major steps before giving detailed information for each step.

Algorithm 9.2 (Major Steps of Piecewise Harmonic Reconstruction).

Input: Discretization of a 3D manifold M (in 3D space) and its boundary $D = \partial M$. Some randomly arranged sample points are located on the boundary surface.

Output: A smooth extension based on sample points on the boundary surface; we also obtain the smooth extension for the whole 3D manifold.

Step 1: D is represented in a simplicial complex form (triangulated format) or cubical form. Sample points are located randomly in D. Find $\{A_1, A_2, \ldots, A_m\}$ in the set of real numbers, R, so that the sample points on D satisfy the gradual variation condition. We could also satisfy the Lipschitz condition with the smallest constant L.

Step 2:   Partition D into simply connected components, so that each sample point is at the edge (boundary) of the partition.

Step 3:   Extract the boundary curve for each component. Each boundary curve is a simple closed curve (path). If there is a pathological case, correct it using an interactive process. Make sure that all sample points are on the boundary curve.

Step 4:   Do a gradually varied interpolation on the boundary curves obtained in Step 3.

Step 5:   Randomly assign values to the inner points of each component, or use gradually varied fitting to get the initial values of the inner points in each component (this is a fast way of converging in the next step).

Step 6:   Harmonic data reconstruction: Update the value of the inner points based on the average value of neighboring points. This process is done for each component until they converge. Add the smooth subdivision algorithm as needed.

Step 7:   3D reconstruction: Use the boundary surface calculated above for all inner points of 3D manifold M. Randomly assign values to each inner point of M.

Step 8:   Harmonic data reconstruction for 3D: Update the value of the inner points based on the average value of neighboring points. This process is done for each component until they converge.

In practice, Steps 4 and 5 can be merged into one step by doing a gradually varied fitting on the whole D, the partitioned boundaries naturally satisfy gradual variation. In addition, Steps 7 and 8 can also be merged into one step by doing a gradually varied fitting on the whole M as direct fitting the M. Or partition M into smaller solid objects. For instance, if $M - \partial D$ contains sample points, then we may need to do a decomposition of M to make sure that each sample point is on the boundary surface of the small solid objects.

Now, we will explain the detailed implementation of each step. The most challenging is Step 2.

Algorithm for Step 2: There will be three sub-algorithms:

Sub-algorithm 1:   The simplest algorithm will be to link all of the points one by one, until they form a cycle. Since we have a finite number of points, using the process would always result in finding such a cycle. If a simple cycle is found that already contains all of the sample points, we use this for the data reconstruction. See the example in the next section.

Otherwise, we start at any point in the fixed cycle. We link this to the sample points until a cycle is formed that contains the rest of the sample points. When, we link the last sample point to the existing cycle, the process will then stop. This is a linear time algorithm using a queue.

Sub-algorithm 2:   Geodesic-like partition. Based on two points that are near each other, find a large cycle. Select a point in the cycle, and select a sample point to find a large cycle. Continue until all sample points are included. The process will then stop.

Sub-algorithm 3:   Delaunay decomposition of D. Each sample point will be on the vertices of triangles. This is a typical case of meshing in computer graphics. So we will not discuss this in detail.

If there is a hole in a partitioned patch, the edge of the hole will be isolated. Therefore, it is not connected to the other edges of the partition. We have the two following lemmas.

Lemma 9.1. *For any closed 2D simply connected discrete surface, if all points on the edge of a partition are connected, then each partitioned patch is simply connected.*

Lemma 9.2. *For any closed 2D discrete surfaces with genus g, if all sample points are connected by gradually varied paths (each vertex in the path has two neighbor vertices in the path), then the paths are the boundaries of several manifolds.*

Based on the theorems provided in Sects. 9.1 and 9.2, we have the harmonic extension for partitioned manifolds.

9.3.3 Case Study: The Algorithm Analysis

In this subsection, we perform some rough algorithm analyses on the algorithms designed in the above subsection.

As we discussed, the input for the main algorithm is: (1) M, the 3D manifold. (2) D, the boundary of M, a closed 2D surface in the discrete format. Let us assume D is triangulated or a set of polygons in 3D (2D cell complexes). (3) J, sample points, which is the subset of D with assigned values in R.

There are two types of distances: geodesic distance of two points on the boundary of M and digital-discrete distance of two points on D, i.e. the graph-distance or the number of vertices between one vertex and another. Geodesic distance describes distance on the edge, if there is an edge in D; otherwise, the shortest path is used to approximate the real geodesic distance. For the remainder of the book, distance refers to graph-distance and not geodesic-distance, unless otherwise specified.

The output of the main algorithm is: a smooth extension based on sample points on the boundary surface; also, the smooth extension for the whole 3D manifold M.

In order to make algorithm analysis simpler, we assume D is a digital manifold. The number of adjacent neighbors of a cell is smaller than a constant. For general case, see [9].

Lemma 9.3. *The time complexity for completing Step 1 in Algorithm 9.2 is* $O(|J||D|log|D|)$.

Proof. Calculate the distance for each pair of points in J. This is exactly like calculating $|J|$ times the Dijkstra algorithm. The Dijkstra algorithm uses $O(|E| + |V|log|V|)$ time [4], where $|E|$ and $|V|$ are the number of edges and vertices. Since $|E| = O(|V|)$ in D, the lemma has been proven. $\square$

We can get both distances types. Next, we calculate the "tangent" or "slope" of each pair of points if we just want to get a Lipschitz extension for the initial values. For non-Lipschitz extensions, we need to use the algorithms described in [2].

Lemma 9.4. *The time complexity to complete Steps 2 and 3 in Algorithm 9.2 is* $O(|D|)$. *If always link to a closer point in cycle finding, this algorithm needs* $O(|D| + |J|^2 \log |J|)$ *for the sorting process for each point in* J.

Proof. Step 2 in Algorithm 9.2 partitions D into simply connected components, so that each sample point is at the edge (boundary) of the partition. Select a point p in J, mark this point then link a line (the shortest path to any other point q in J. If the link contains a marked point, there will be a cycle. Otherwise, just mark q, select new one r and link q to r. So this algorithm always mark only once at a point. This algorithm is linear. (Assume the shortest path is known for each pair.)

If we always find a closest unmarked point in J, we need $O(|J|^2 \log |J|)$ for the sorting process for each point in J. According to Step 2, we can use geodesic distance to find a line from one point to another point in J. This way eliminates some of the paths that are the same digital-discrete distance. The algorithm needs to find two different lines from each point visited and this is a linear algorithm of $|D|$ since it is possible that every point in D will be included. This is because we obtain the distance (weighted) and fill the paths to each. When a point is visited, we mark it in a queue and continue in order to get the objective points. When a visited point already has more than two neighbors that have been visited, we delete the next point from the queue. $\square$

This process can be combined with Step 4 of Algorithm 9.2. As we discussed, Steps 4 and 5 can be merged into one step by doing a gradually varied fitting of the whole D. The function on the partitioned boundaries naturally satisfy the gradual variation.

Lemma 9.5. *The time complexity for completing Steps 4 and 5 of Algorithm 9.2 is* $O(|D||D|)$. *The improved algorithm can reach* $O(|D|\log|D|)$.

Proof. This step assigns values to all the marked points using the gradually varied fitting algorithm. It usually takes $O(|D|^2)$ if a simple algorithm is implemented.

This algorithm can be implemented in $O(|D|\log|D|)$ since the decomposition of the mesh is Jordan [6, 7]. $\square$

Lemma 9.6. *The time complexity for completing Step 6 of Algorithm 3.1 is* $O(|D|^3)$ *if we use the Gaussian elimination method. It can usually be done in* $O(|D|\log|D|)$ *if we use fast algorithms for sparse symmetrically diagonally dominant linear systems.*

Proof. There are two kinds of implementations for iteration or solving the system of linear equations. (1) Randomly assign values to the inner points of each component, or use gradually varied fitting to get the initial value of the inner points of each component (a fast way to converge for the next step).

(2) This stage can also be implemented by solving a system of linear equations. The linear equations can be formed in the following way:

All unknown vertices are named $x_1, \cdots, x_N$. For each x_i, we get its neighbors, some of which are in $x_1, \cdots, x_N$, and the rest are known (meaning that we know their values). According to the definition of the equation for discrete harmonic functions, we have:

$$a_{i1}x_1 + a_{i2}x_2 + \cdots + a_{ii}x_i + \cdots + a_{iN}x_N = C_i \tag{9.7}$$

where $a_i i$ equals to the integer indicating the number of neighbors. a_{ik} is either equal to "−1" or "0," where "−1" means that x_k is a neighbor of x_i. C_i is a real number that is equal to the summation of the values of neighboring points not included in the set $x_1, \cdots, x_N$.

In most cases, the number of neighbors is bounded. See the discussions in Sect. 9.2. The matrix of coefficients a_{ij} is symmetric since if a_{ik}, k is not equal to i and is equal to "−1," which means x_k is a neighbor of x_i. Therefore, x_i is a neighbor of x_k, that is to say a_{ki} is also "−1."

If we use a cubic space in 3D, there will only be 6–26 non-zeros in a row since a voxel has at most 26 adjacent voxels.

This system of linear equations is a sparse symmetrically diagonally dominant linear system, usually with a constant band-width in the diagonal, which means that there is a constant number of non-zero elements in a row [15, 16]. The recent result by Koutis et al. has reached the optimal time to get the solution $O(mlognlog(1/epsilon))$, where $epsilon$ is a given constant for accuracy measurement [16]. □

Lemma 9.7. *The time complexity for completing Steps 7 and 8 of Algorithm 9.2 is $O(|D|^3)$ if we use the Gaussian elimination method. It can usually be done in $O(|D|log|D|)$ if we use a fast algorithm for sparse symmetrically diagonally dominant linear systems.*

Proof. The proof is the same as the one for Lemma 3.6. We implement the same algorithm to fill the inner points of a 3D manifold with the values on the boundary surfaces. This algorithm does not need to assume the convexity of a geometric object. □

Therefore, we have the following theorems.

Theorem 9.5. *There is a near $O(|D|log|D|)$ algorithm (Steps 1–6 of Algorithm 9.2) for a Lipschitz or gradually varied based piecewise harmonic function extension on a 2D closed surface.*

Theorem 9.6. *There is a near $O(|D|log|D|)$ algorithm for a 3D harmonic fill where the boundary surface is a piecewise harmonic function.*

9.3.4 Open Problems

More sophisticated algorithms can be done in future research. We have three open problems listed below.

Open Problems:

1. The Balanced partition: each component is about the same as the sample size. Is the time complexity P or NP-hard?
2. The minimum length partition: The length of all curves is minimal. This problem in the cubical case might not be NP-hard. Why?

3. The largest number of inner points (similar to the second problem): There are more smooth parts in the fitted function. Is this problem NP-hard?

9.4 Implementation and Examples

The implementation is based on the basic property of discrete harmonic functions for 2D closed manifolds. A closed curve partition is needed to get an appropriate function extension. As long as the function on the curve (path) is continuous (gradually varied), the inside of the curve will be uniquely defined based on the harmonic function.

As we discussed many times, data reconstruction is used to fit a function based on the observations of some sample (guiding) points. For instance, we have five sample points on a sphere. We can use a normal line at the point where the height of the line indicates the value of the sample point. We can link all the sample points to form a closed curve, which is piecewise linear. Its corresponding path on the sphere separates the domain into two regions or components.

Even though we prefer to use piecewise harmonic extension, we can still use harmonic fitting for randomly selected data points. Figure 9.6(a), (b) also shows some similarities to directly fitting based on discrete sample points (not a path). More research should be done to determine the conditions in which we can directly use a harmonic function extension without going through the piecewise harmonic method.

In the example in Fig. 9.7(a), when multiple cycles are constructed for surface decomposition, we can still get a very good harmonic fitting. See Fig. 9.7(b).

The last example shows the guiding points that are not connected initially. However, our algorithm will find the paths connecting these points. There are 12 guiding points in this example. The algorithm will connect all guiding points into paths; those paths will make a partition to the manifold. If one fit the paths first, then the harmonic reconstruction can be based on the fitted paths is shown in Fig. 9.8.

Figure 9.9 shows the automatic partition using our Algorithm 9.2 (Sub-Algorithm 1).

The final reconstruction result is shown in Fig. 9.10.

9.5 Further Discussions

Our new method presented in [9] contains a philosophical change in computer graphics. Triangulation is no longer necessary with this new method, a discovery which is incredibly exciting! Moreover, the method we present is the generalization of triangulation and bi-linear interpolation (in Coons surfaces, this refers to four point interpolation). Triangulation and bilinear methods are both harmonic. However, harmonic functions do not only refer to linear functions and they are much more general.

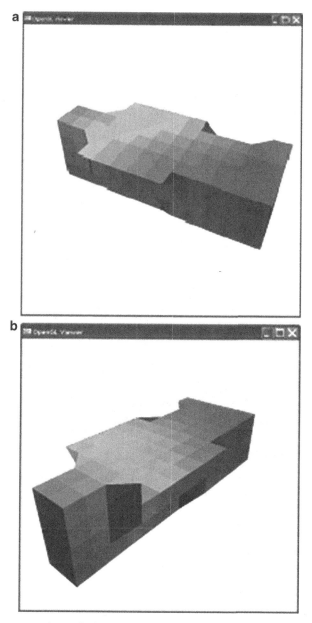

Fig. 9.6 Using seven points to fit the data on a 3D surface: (**a**) GVF result. (**b**) Harmonic fitting (with a few iterations) based on GVF

The classical discrete method for data reconstruction is based on domain decomposition according to guiding (or sample) points. Then, the spline method (for polynomials) or finite elements method (for PDE) is used to fit the data. Our method

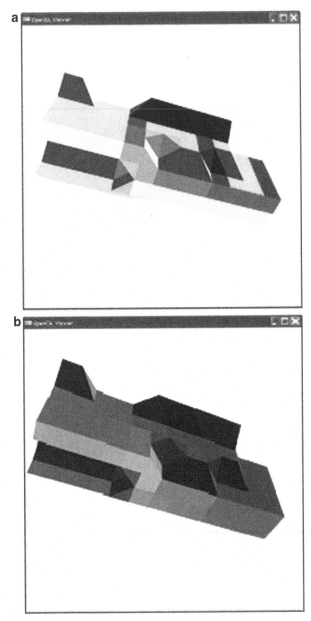

Fig. 9.7 Three components of the 2D manifolds: (**a**) GVF on the paths. (**b**) Harmonic fitting using GVF as guiding points

is based on a gradually varied function that does not assume the linearly separable property among guiding points, i.e. no domain decomposition methods are needed.

Fig. 9.8 An example with 12 sample points from difference views

We also demonstrate the flexibility of the new method and its potential in solving a variety of problems.

The approach presented in this chapter is a new treatment to the important problem in the applications of numerical computation and computer graphics. The following are some popular cited research for data reconstruction. Levin used the moving least squares (MLS) method for mesh-free surface interpolation [17, 18]. This method still requires the use of dense sample points, meaning that if there are not enough points in the neighborhood then the size of the neighborhood must be adjusted. This is not an automated solution. An adaptive process is needed, which requires an artificial intelligence resolution. Our method tries to find the first fit for MLS or other subdivision methods. We discuss some possible research topics in Chap. 12.

Bajaj et al. used Delaunay triangulations for automated reconstruction of 3D objects [1]. Again, this uses the linear interpolation as the first treatment. Harmonic functions, on the other hand, are not limited to linear functions. In fact, all linear

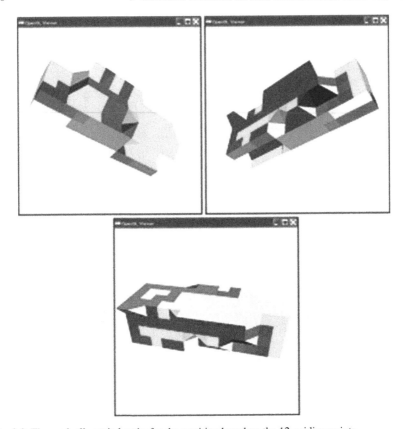

Fig. 9.9 The gradually varied paths for the partition based on the 12 guiding points

functions are harmonic. Therefore, harmonic functions can use a polygon as the boundary and is not limited to only triangles and rectangles.

The difference between gradually varied fitting and harmonic fitting is presented in Chap. 11.

Acknowledgements Professor Thomas Funkhouser at Princeton University provided helps on the 3D data sets and OpenGL display programs. Professor Zhihua Zhang at Zhejiang University has provided references on the irreducibly diagonally dominant. The author would like to express thanks to them.

References

1. Bajaj CL, Bernadini F, Xu G (1995) Automatic reconstruction of surfaces and scalar fields from 3D scans. In: Proceedings of the SIGGRAPH, Los Angeles, pp 109–118
2. Benjamini I, Iovasz L (2003) Harmonic and analytic frunctions on graphs. J Geom 76:3–15
3. Catmull E, Clark J (1978) Recursively generated B-spline surfaces on arbitrary topological meshes. Comput Aided Des 10(6):350–355

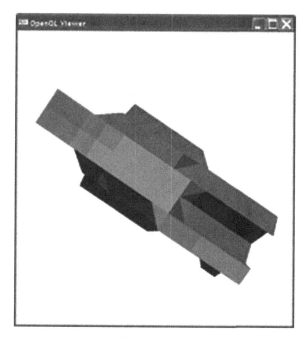

Fig. 9.10 The harmonic reconstruction of the surface on a 2D manifold based on the reconstructed paths

4. Chen L (2010) Digital-discrete surface reconstruction: a true universal and nonlinear method. http://arxiv.org/ftp/arxiv/papers/1003/1003.2242.pdf
5. Chen L (2004) Discrete surfaces and manifolds. Scientific and practical computing, Rockville
6. Chen L (2005) Gradually varied surfaces and gradually varied functions, in English 2005; in Chinese, 1990 CITR-TR 156, University of Auckland
7. Chen L (2010) A digital-discrete method for smooth-continuous data reconstruction. J Wash Acad Sci 96(2):47–65. ISSN 0043–0439, Capital Science 2010 of The Washington Academy of Sciences and its Affiliates, March 27–28, 2010
8. Chen L, Liu Y, Luo F (2009) A note on gradually varied functions and harmonic functions. http://arxiv.org/PS_cache/arxiv/pdf/0910/0910.5040v1.pdf
9. Chen L, Luo F (2011) Harmonic Functions for Data Reconstruction on 3D Manifolds, Submitted for publication, http://arxiv.org/ftp/arxiv/papers/1102/1102.0200.pdf
10. Courant R (1950) Dirichlet's principle, conformal mapping, and minimal surfaces (appendix: Schiffer M), Interscience, New York
11. Demmel JW (1997) Applied numerical linear algebra. SIAM, Philadelphia
12. Gilbarg D, Trudinger N (2001) Elliptic partial differential equations of second order. Classics in mathematics. Springer, Berlin
13. Golub GH, Van Loan CF (1996) Matrix computations, 3rd edn. The Johns Hopkins University Press, Baltimore
14. Heilbronn HA (1949) On discrete harmonic functions. Math Proc Camb Philos Soc 45:194–206
15. Koutis I, Miller GL, Peng R (2010) Approaching optimality for solving SDD systems. CoRR, *abs*1003.2958
16. Koutis I, Miller GL, Peng R (2011) Solving SDD linear systems in time $O(mlognlog(1/epsilon))$. http://arxiv.org/PS_cache/arxiv/pdf/1102/1102.4842v3.pdf

17. Lancaster P, Salkauskas K (1981) Surfaces generated by moving least squares methods. Math Comput 87:141–158
18. Levin D (2003) Mesh-independent surface interpolation. In: Brunnett G, Hamann B, Miller H (eds) Geometric modeling for scientific visualization. Springer, Berlin
19. Varga RS (2002) Matrix iterative analysis, 2nd edn. (of 1962 Prentice Hall edition). Springer, New York
20. Varga RS (2004) Gersgorin and his circles. Springer, Berlin

Part III
Advanced Topics

Chapter 10
Gradual Variations and Partial Differential Equations

Abstract Numerically solving partial differential equations (PDE) has made a significant impact on science and engineering since the last century. The finite difference method and finite elements method are two main methods for numerical PDEs. This chapter presents a new method that uses gradually varied functions to solve partial differential equations, specifically in groundwater flow equations. In this chapter, we first introduce basic partial differential equations including elliptic, parabolic, and hyperbolic differential equations and the brief descriptions of their numerical solving methods. Second, we establish a mathematical model based on gradually varied functions for parabolic differential equations, then we use this method for groundwater data reconstruction. This model can also be used to solve elliptic differential equations. Lastly, we present a case study for solving hyperbolic differential equations using our new method.

10.1 Numerical Solutions for Partial Differential Equations

A partial differential equation (PDE) is an equation that relates a function to its partial derivatives [10, 12]. Since first-order partial differential equations (involving only first order derivatives) can be solved by integration, in this section, we give very brief introductions to the most common types of second order partial differential equations. Then, we present the finite difference method for solving these types of equations numerically.

10.1.1 Three Types of Partial Differential Equations

The general form of the second-order partial differential equations is

$$Au_{xx} + Bu_{xy} + Cu_{yy} + Du_x + Eu_y + F = 0 \qquad (10.1)$$

where $u_{xx} = \frac{\partial^2 u}{\partial x^2}$, $u_{xy} = \frac{\partial^2 u}{\partial x \partial y}$, and $u_{yy} = \frac{\partial^2 u}{\partial y^2}$.

L.M. Chen, *Digital Functions and Data Reconstruction: Digital-Discrete Methods*,
DOI 10.1007/978-1-4614-5638-4_10, © Springer Science+Business Media, LLC 2013

The solution for the equation performs differently based on the values of A, B, and C. We can classify the different equations into parabolic, elliptic, and hyperbolic based on the values of $B^2 - 4AC$, which is called the discriminant.

(a) $B^2 - 4AC < 0$ is elliptic. One example is Laplace's equation , which we discussed in Chap. 9. The Poisson equation is more general:

$$\frac{\partial^2 u}{\partial x^2} + \frac{\partial^2 u}{\partial y^2} = f(x,y), \qquad (10.2)$$

or for simplicity:

$$\Delta u = \nabla^2 = f$$

where Δ is called the Laplace operator.

(b) $B^2 - 4AC = 0$ is parabolic. An example is the heat equation , which is used when heat transfers in a media.

$$\frac{\partial u}{\partial t} = c\left(\frac{\partial^2 u}{\partial x^2} + \frac{\partial^2 u}{\partial y^2}\right) \qquad (10.3)$$

(c) $B^2 - 4AC > 0$ is hyperbolic, the solution for this type of function is much different than those of elliptic and parabolic functions. Stability is the main concern for these numerical solutions. An example is the wave equation.

$$\frac{\partial^2 u}{\partial^2 t} = c\left(\frac{\partial^2 u}{\partial x^2} + \frac{\partial^2 u}{\partial y^2}\right) \qquad (10.4)$$

These three types of partial differential equations are usually called mathematical physics equations. Some of these equations do not have an explicit solution. In engineering, we usually look for numerical solutions . Two popular methods for PDE are the finite difference method and the finite elements method. In this section, we introduce the finite difference method.

10.1.2 The Finite Difference Method for PDE

The finite difference method can be used to solve all three types of equations. The solving processes for the three types is the same: decompose the domain into a 2D grid (2D digital space), use difference formulas to represent the differentials at each grid point, and use the system of linear equations to solve [13, 16]. This is possible because the PDE will generate a relation at the grid point and this relation is a linear equation. We discussed harmonic equations in Chap. 9, which has a similar form.

Since each unknown point $u(x,y)$ has a value that has yet to be determined, the grid space generates a new linear equation where the number of unknown points is the same as the number of equations. In general, the solution for the system of linear equations is unique if the sample point set is not empty. The "sample points" are usually the boundary values or initial values. When a time variable t is involved in the equation, we usually fit the first function or first two functions, $t = 0, 1$, as the initial surfaces before doing other fittings.

The following are the three most common finite difference equations for these differential equations.

According to the finite difference format of second order derivative (Chap. 2) we have $u_{i,j} = u(x_i, y_j)$, $x_i = x_0 + i\Delta$ and $y_j = y_0 + j\Delta$.

Without loss of generality, we always use the central difference scheme to discretize functions.

$$u_{xx} = \frac{u_{i+1,j}^k - 2u_{i,j}^k + u_{i-1,j}^k}{(\Delta)^2} \text{ and } u_{yy} = \frac{u_{i,j+1}^k - 2u_{i,j}^k + u_{i,j-1}^k}{(\Delta)^2} \tag{10.5}$$

It is straightforward to use the finite difference form instead of the differential form in obtaining the finite difference equations. We list three of them.

(a) The Poisson equation's finite difference format is as follows. According to the finite difference format of second order derivatives, we have,

$$(u_{i+1,j} + u_{i-1,j} + u_{i,j+1} + u_{i,j-1} - 4u_{i,j}) = \Delta^2 f_{i,j} \tag{10.6}$$

It is the same as the formula presented in Chap. 9.

(b) The heat equation's finite difference form has two parts: the left side is related to time t and the right side is a Laplace function. The original heat differential equation is

$$u_t = c \cdot (u_{xx} + u_{yy})$$

For the left side, we use u^k to represent function u at time k, so

$$u_t(x, y) = \frac{u^{k+1}(x, y) - u^k(x, y)}{\Delta t} \tag{10.7}$$

The right side is

$$\frac{u_{i,j}^{k+1} - u_{i,j}^k}{\Delta t} = c \cdot \left(\frac{u_{i+1,j}^k - 2u_{i,j}^k + u_{i-1,j}^k}{(\Delta)^2} + \frac{u_{i,j+1}^k - 2u_{i,j}^k + u_{i,j-1}^k}{(\Delta)^2} \right). \tag{10.8}$$

To express $u_{i,j}^{k+1}$ explicitly,

$$u_{i,j}^{k+1} = u_{i,j}^k + c \cdot \frac{\Delta t}{(\Delta)^2} \left(u_{i+1,j}^k + u_{i-1,j}^k + u_{i,j+1}^k + u_{i,j-1}^k - 4u_{i,j}^k \right). \tag{10.9}$$

This equation is used in groundwater equations for practice in a later section. Note that this equation has a stability region when we apply the difference equation mentioned above. This means that if the condition is under the stability condition, the result can be trusted [7, 13]. The stability condition requires

$$r = c \cdot \frac{\Delta t}{(\Delta)^2} \leq \frac{1}{2}.$$

In real data processes, Δt and/or Δ may be set to 1, which would make the problem a digital space problem. $c \leq \frac{1}{2}$ means that computationally, $u_{i,j}^k$ adds many more contributions to $u_{i,j}^{k+1}$. Using $\Delta u = (u^{k+1} - u^k) \cdot \Delta t$ as the derivative at time k is called forward difference. If we use $\Delta u = (u^k - u^{k-1}) \cdot \Delta t$, we will have so called backward difference. The difference formula for the heat equation is:

$$u_{i,j}^k = u_{i,j}^{k-1} + c \cdot \frac{\Delta t}{(\Delta)^2} \left(u_{i+1,j}^k + u_{i-1,j}^k + u_{i,j+1}^k + u_{i,j-1}^k - 4u_{i,j}^k \right). \tag{10.10}$$

This is called an implicit form. It is also stable in almost any condition [7, 16].

(c) The original wave equation is

$$u_{tt} = v^2 \cdot (u_{xx} + u_{yy}).$$

For the left difference form, we use u^k to represent function u at time k. So the left part of the wave equation in the finite difference format is

$$u_{tt}^k(x_i, y_j) = \frac{u_{i,j}^{k+1} - 2u_{i,j}^k + u_{i,j}^{k-1}}{(\Delta t)^2} \tag{10.11}$$

The right part is

$$\frac{u_{i,j}^{k+1} - 2u_{i,j}^k + u_{i,j}^{k-1}}{(\Delta t)^2} = v^2 \cdot \left(\frac{u_{i+1,j}^k - 2u_{i,j}^k + u_{i-1,j}^k}{(\Delta)^2} + \frac{u_{i,j+1}^k - 2u_{i,j}^k + u_{i,j-1}^k}{(\Delta)^2} \right). \tag{10.12}$$

Therefore we have:

$$u_{i,j}^{k+1} = 2u_{i,j}^k - u_{i,j}^{k-1} + \frac{v^2(\Delta t)^2}{(\Delta)^2} \left(u_{i+1,j}^k + u_{i-1,j}^k + u_{i,j+1}^k + u_{i,j-1}^k - 4u_{i,j}^k \right). \tag{10.13}$$

The stability requirements of this difference is

$$r = v^2 \cdot \frac{\Delta t}{(\Delta)^2} \leq \frac{1}{2}$$

For wave equations, a simple explanation of stability is the following: $\frac{\partial^2 u}{\partial^2 t} = c\Delta u(x, y)$, where c is related to the speed of the wave. When sampling Δx and Δy, if these two points are small, then Δt must also be small since the two sample points may belong to different cycles of the wave. For example if the two sample points are selected at two peaks, then the reconstruction will not contain details of the valleys. Therefore, the stability condition in 1D, for easy understanding, is

$$\frac{|c|\Delta t}{\Delta x} \leq 1$$

The finite difference method is meant to directly get the function value. We discuss the finite elements method in Chap. 11.

10.2 Digital-Discrete Methods for Partial Differential Equations

The finite difference method requires strict boundary values in order to find a valid solution in the region. This is because the solution inside the region is smooth. The difficulty of real world problems is that in many cases, we have sample data points inside the domain and not on the boundary.

Another problem is that c or v in the above equations (in addition to stability conditions) are not always constant in the real world. The partial differential equations might not have a unique solution using the system of linear equations.

Digital-discrete methods can provide a reasonable answer for these partial differential equations. The key idea is to use an iterated process to approximate the solution.

In addition, especially for the initial condition ($t = 0$), the digital-discrete method can fit the function first as described in Chaps. 7–9, then use the function to extend to the next time point. In Chap. 11, we will explain why the digital-discrete method has the same flexibility as the finite elements method, but is much faster and easier in implementation. Future research could include incorporating the least squares method.

The general strategies for solving problems using digital-discrete methods are as follows: Do an initial gradual varied fitting and modify the fitted function based on the specific partial differential equations (instead of using Taylor expansion discussed in Chap. 7). The modification can be iterated.

10.2.1 Digital-Discrete Algorithm for Poisson Equations

Poisson's equation is a type of elliptical PDE. We can also use the same method to solve other elliptical PDEs. According to the finite difference format of second order derivatives (10.6), we have the finite difference equation for Poisson's equation.

$$(u_{i+1,j} + u_{i-1,j} + u_{i,j+1} + u_{i,j-1} - 4u_{ij}) = \Delta^2 f_{i,j} \qquad (10.14)$$

Let us assume that f_{ij} is known for some of the equations, or at least cover of the sample points (more than the sample points).

The easiest process would be to fix three points (using the GVF value as the initial value for the three points) in order to solve for one point, and then use the new value for the point to update the old GVF values. We expand the sample point set in this way until the entire domain is fitted. If $f_{i,j}$ is not known for the whole area, then we also need to use the equation to predict the unknown f value. The initial value for the unknown f can be obtained by doing a gradually varied fitting of f.

Let u_{ij} be known for the sample points. Selecting all equations that contain u_{ij}, we then update the value in a subdomain (a small disk) near the sample points, before expanding the unknown data points to the entire area. We use the GVF fitted function as reference data to iterate the sample points from the center of the function in order to update their values.

For instance, if there is only one sample point $u_{i,j}$ in a small disk region, we have four variables that need to be solved $u_{i+1,j}$, $u_{i-1,j}$, $u_{i,j+1}$, and $u_{i,j-1}$. Using a relatively bigger circle containing five points, we randomly select another three equations. Then, the GVF fitted value will replace all data that is not one of our four variables $u_{i+1,j}$, $u_{i-1,j}$, $u_{i,j+1}$, or $u_{i,j-1}$; find the solution; and then let $g(x,y) = gvf(x,y)$, where we use $(u_{x,y} + g(x,y))/2$ initially to get an updated value for the new $g(x,y)$.

A more general procedure can be designed as follows: For any subdomain A, where ∂A is the boundary of A, a set of sample points in $A - \partial A$ can be considered as sample points in A. Use the GVF fitted values on ∂A to update the values in $A - \partial A$. The size of A should be kept relatively small. When two or more disks cover one special point, we can use the average value.

Since we would have more linear equations than variables if a larger disk is applied, we could apply the least squares method to solve the equations in order to get new disks u around the sample points. When every point has been fitted, the iteration could continue until there is convergence.

10.2.2 Digital-Discrete Algorithm for the Heat Equation

The heat or diffusion equation was presented above in its finite difference form. The difference equations have a unique solution when the sample data is on the boundary [13] (in addition to the initial conditions). The problem is the same as the one discussed in Sect. 10.2.1 where the sample points are not usually on the boundary. If we are required to use the difference equations directly, we would get a solution that is not smooth at the sample points. Therefore, the results using this method are not the solutions to our question.

The key is how we transfer the current data function to the next time point. For simplicity, we assume $\Delta t = \Delta = 1$.

We can first solve:

$$u_{t=0} = c \cdot (u_{xx} + u_{yy})$$

using the algorithm for Poisson Equations in Sect. 10.2.1 as the initial surface. Then, we can use the following equation to update the time $t > 0$.

$$u_{i,j}^{t+1} = u_{i,j}^t + c \cdot \left(u_{i+1,j}^t + u_{i-1,j}^t + u_{i,j+1}^t + u_{i,j-1}^t - 4u_{i,j}^t \right). \tag{10.15}$$

When (i,j) is a sample point, $u_{i,j}^t$ may be known for all t. We only need to update $u_{i,j}^{t+1}$ for the unknown locations.

After obtaining the digital-discrete fitting at $t+1$, update the time $t+1$ function with the iteration. For example, use $u_{i,j}^{t+1} \leftarrow (u_{i,j}^{t+1} + g(i,j))/2$ (it is also possible to update $u_{i,j}^k$, $k = 0, \cdots, t+1$). This algorithm is used in groundwater applications in this chapter, Sect. 10.3.

10.2.3 Digital-Discrete Algorithm for the Wave Equation

The digital-discrete method for wave equations are similar to that of the heat equation. The difference calculating the first two time functions at $t = 0, 1$ according to the initial conditions $u_{t=0} = f_0$ and $\frac{du}{dt}|_{t=0} = f_1$.

$$u_{i,j}^{t+1} = 2u_{i,j}^t - u_{i,j}^{t-1} + \left(u_{i+1,j}^t + u_{i-1,j}^t + u_{i,j+1}^t + u_{i,j-1}^t - 4u_{i,j}^t\right). \qquad (10.16)$$

Using a digital-discrete fitting at $t+1$, update the time $t+1$ function with the iteration. For example, use $u_{i,j}^{t+1} \leftarrow (u_{i,j}^{t+1} + g(i,j))/2$.

10.3 Case Study: Groundwater Flow Equations

Research in groundwater flow is essential since it impacts our daily lives [1, 8, 9, 11, 15, 17]. The groundwater flow equation is a diffusion equation, which is a type of parabolic equation. It is based on Darcy's law and usually describes the movement of groundwater in a porous medium, such as aquifers.

In terms of groundwater, the Laplace equation is an estimation of a steady-state flow and is a simplification of the diffusion equation under the condition that the aquifer has a recharging and/or discharging boundary.

The current method of flow estimation mainly uses the groundwater flow equation. Computer source codes such as MODFLOW solve 2D equations and pass the data vertically to create 3D volume [9, 11]. Much research has already been done to find a discrete model for the groundwater flow equations.

The research is mainly based on numerical and analytical methods [8, 15, 17]. The finite difference method (FDM) and the finite elements method (FEM) are popular in this area. Pruist et al. [14] indicated that FEM has advantages in local refinements of grid (adaptive mesh generation) due to non-rectangular grids, increased accuracy, stability, and representation of the spatial variation of anisotropy. However, its computational cost is much larger and is relatively less manageable in application. In fact, the groundwater industry is not like the automobile industry in that there is not much need for a good-looking smoothed groundwater level surface.

On the other hand, FDM has simplicity in terms of theory and algorithms, in addition to being easy in application; however, FDM has problems regarding inefficient refinement of grids and poor geometric representations. This is due to the use of strictly rectangular grids. Also, there is no standard method of implementing the Neumann boundary condition.

The gradually varied function is supposed to pick up the advantages and overcome the disadvantages of FEM and FDM. So our first task is to investigate how suitable gradually varied functions are for groundwater data. Then, we must find a connection between the flow equations and gradually varied functions. We also need to design an input data format to store the data in a database.

A gradually varied function is for a discrete system where a high level of smoothness is not the dominant factor. It can be used in any type of decomposition of the domain. This method is more flexible than rectangle-cells used in MODFLOW and triangle-cells used in FEFLOW, a software based on the finite elements method. Because gradual variation does not have strict system requirements, other mathematical methods and the artificial intelligence method can easily be incorporated into this method in order to seek a better solution. Based on the boundary conditions or constraints of the groundwater aquifer, the constraints could be in explicit form or differential form, such as diffusion equations. The gradually varied function exists based on the following theorem: The necessary and sufficient condition of the existence of a gradually varied function is that the change of values in any pair of sample points is smaller than or equal to the distance between the pair of points [2, 5].

10.3.1 Background and Related Research

The groundwater flow equation based on Darcy's law usually describes the movement of groundwater in a porous medium such as aquifers. Conservation of mass equation, for a given increment of time Δt), is given as follows:

$$\frac{\Delta W_{store}}{\Delta t} = \frac{W_{in}}{\Delta t} - \frac{W_{out}}{\Delta t} - \frac{W_{generate}}{\Delta t} \qquad (10.17)$$

W_{store} indicates the water in the aquifer, W_{in} and W_{out} indicate water flowing-in and out across boundaries, and $W_{generate}$ is indicates the volume of the source within the aquifer. Its differential form is

$$\frac{\partial h}{\partial t} = \alpha \left[\frac{\partial^2 h}{\partial x^2} + \frac{\partial^2 h}{\partial y^2} + \frac{\partial^2 h}{\partial z^2} \right] - G \qquad (10.18)$$

where h is the hydraulic head, a mathematical term consisting of the function's aquifer storage coefficient S, aquifer thickness b, and aquifer hydraulic conductivity K. G is the sink/source term such as groundwater pump, recharge, etc. To solve Eq. 10.18, a grid method is usually used, such as the finite difference or finite elements method [8, 9, 11, 17]. Other methods, including the analytic element method, attempt to solve the equation exactly, but require approximations of the boundary conditions [8, 15]; they are mainly used in academic and research labs.

The most popular system is called MODFLOW, which was developed by USGS [11]. MODFLOW is based on the finite-difference method on rectangular Cartesian coordinates. MODFLOW can be viewed as a "quasi 3D" simulation since it only deals with the vertical average (no z-direction derivative). Flow calculations between the 2D horizontal layers use the concept of leakage.

To use gradually varied functions and digital-discrete methods to solve this problem is to use Darcy's law or differential constraints to determine the hydraulic head value of the unknown points instead of using a random selection of constructed gradually varied surfaces when there is more than one possible selection [5]. When the determinations of the values are uncertain, we use random fitting algorithms in the process [3, 5].

The method described in Chap. 7 can be used for a single surface fitting if the condition in the theorem is satisfied. The problem is when sample data does not satisfy the condition of fitting and the original algorithm cannot be used directly for individual surface fitting.

10.3.2 Parabolic Equations and Digital-Discrete Solutions

The digital-discrete method overcomes the problem of gradually varied surfaces. The digital-discrete method is able to obtain a smooth surface. The idea behind using a digital-discrete method to solve the groundwater equation is to fit one individual surface at a time, then use the difference method to update the flow equation.

Algorithm 10.1 (The major steps of the algorithm [4, 6]). Start with a particular dataset consisting of guiding points defined on a 2d space. The major steps of the new algorithm are as follows (this is for 2D functions, we would only need to add a dimension for 3D functions):

Step 1:   Load guiding points. In this step we load the data points with observation values.

Step 2:   Determine the resolution. Locate the points in grid space.

Step 3:   Function extension. This way, we obtain gradually varied or near gradually varied (continuous) functions. In this step, the local Lipschitz condition may be used.

Step 4:   Use the finite difference method to calculate partial derivatives, then obtain the smoothed function. This function is an base function, the first water head surface for the time sequence.

Step 5:   Use the groundwater difference equation (e.g. (10.18), (10.10), and (10.15)) to obtain each surface function at time t.

Step 6:   Some multilevel and multi-resolution methods may be used to do the fitting when the data set is large.

the rest of the subsection will provide details for the real calculation.

10.3.2.1 Individual Surface Fitting

There are three algorithms we have designed to fit the surface at time t for the groundwater equation: (1) A gradually varied surface, i.e. directly using Algorithm 3.1 to fit the data would result in a continuous surface. (2) A smooth gradually varied surface, in other words using Algorithm 7.1 to get a C^2 function. (3) Getting a Poisson solution using the method described in Sect. 10.2.1 after a gradually varied fitting.

10.3.2.2 Sequential Surface Fitting and Involvement of the Flow Equation

An individual surface fitting is not enough since the relationship between groundwater surfaces at a given time point is not accounted for in the calculation. In particular, we have to use the flow equation in the sequential surface calculation.

Using flow Eqs. (10.18) and (10.10), we have a backward difference equation, which would give a stable solution (it is unconditionally stable).

$$h^t_{new}(x,y) - h^{t-1}(x,y) = \alpha(h^t(x-1,y) + h^t(x+1,y) + h^t(x,y-1) + \\ h^t(x,y+1) - 4h^t(x,y)) - G \tag{10.19}$$

where h^t uses a gradually fitted function. We can let

$$f_4 = (h^t(x,y) - h^{t-1}(x,y) + G)/\alpha + 4h^t(x,y) \\ = h^t(x-1,y) + h^t(x+1,y) + h^t(x,y-1) + h^t(x,y+1). \tag{10.20}$$

$f4$ can also be viewed as the average of four times the h values at the center point.

10.3.3 Experiments and Real Data Processing

We have applied the above algorithms to three sets of problems. The first set is data from a region in Northern Virginia. The second set is experimental data around a pumping well. The third set is also data around a pumping well but is real data.

10.3.3.1 Lab Data Reconstruction

In this data set, we have simulated a pumping well. We select 40 observations with locations surrounding the well to get the log data. These points are selected along with eight transects surrounding the pumping well. Five observation points are chosen in each transect [6].

This example shows that our algorithm worked very well on data reconstruction. The correctness of our algorithm is partially verified. If 16 directions with 80 sample points are chosen, the resulting picture will be much circular.

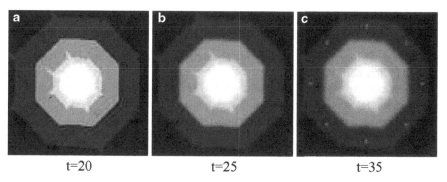

Fig. 10.1 Eight data points at different time points (**a**) t = 20 (**b**) t = 25 (**c**) t = 35

10.3.3.2 Real Data Reconstruction Based on One Pumping Well

This example contains nine samples of real data points for a pumping well [6] (Fig. 10.2).

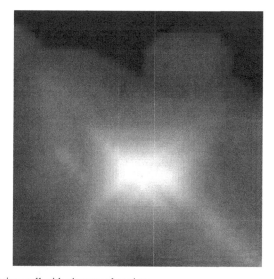

Fig. 10.2 A pumping well with nine sample points

The data distribution has a rectangular pattern. It is caused by the reconstruction process done based on rectangular cells. When there are few sample points and they are not selected randomly, the data will appear to be rectangular in shape.

10.3.4 Real Data Reconstruction for a Large Region

The next experiment uses our method for a relatively large region. This region is located in Southern Virginia.

To test the correctness of the results, we located the data points in the rectangular area on GoogleMaps using the latitude and longitude.

We also tested the following fitted points to make sure they correspond with their location. The selected points are used in reconstruction.

We can find the points at

Table 10.1 Testing data location

Value	Latitude	Longitude
4.65	36.62074879	−76.10938540
75.37	36.92515020	−77.17746768
6.00	36.69104276	−76.00948530
175.80	36.78431615	−76.64328700
168.33	36.80403855	−76.73495750
157.71	36.85931567	−76.58634110
208.26	36.68320624	−76.91329390
7.26	36.78737704	−76.05153760

The data area covers the coast line and some mountainous areas. The groundwater levels in the mountains will be higher and the levels near the sea will be much lower.

Fig. 10.3 The map and ground water data

Figure 10.3 shows a good match between the ground water data and the region's geographical map. The brightness of the pixels mean the deeper the distance from the surface. In mountain areas, the groundwater level is generally lower. However some mismatches may occur due to not having enough sample data points (observation wells).

In order to compare the results from the direct fitting and solving the groundwater equation, we first show the results that were produced using an individualized-fit algorithm to fit the initial data set (see Fig. 10.4). This algorithm is also made by the rough gradually varied surface fitting by scanning through the fitting array. There are many clear boundary lines in the images.

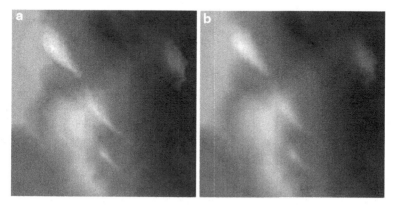

Fig. 10.4 Groundwater header calculation: (**a**) Gradually varied functions for initial surface reconstruction, (**b**) PDE solution for a certain number of iterations

Figure 10.5 shows the sequential surface fitting and updates on the water flow equation. Individual surface fitting results are shown in (a)–(c). Starting with one fitted surface at a time, the process will converge faster. This will not affect the final result if there are enough iterations.

10.4 Future Remarks

To ensure the accuracy of the calculation, we shall add the finite elements method to our research. We also want to use MODEFLOW to calculate local and small-region flow, in addition to using gradual variation to compute regional or global data.

Acknowledgements This research was supported by USGS seed grants. The author expresses thanks to Dr. William Hare and members of the UDC DC Water Resources Research Institute for their help. The lab data was provided by Professor Xunhong Chen at UNL. The author would also like to thank his student Mr. Travis L. Branham for his work in data collection. This chapter is based on the report entitled: "Gradual Variation Analysis for Groundwater Flow of DC, DC Water Resources Research Institute Final Report 2009".

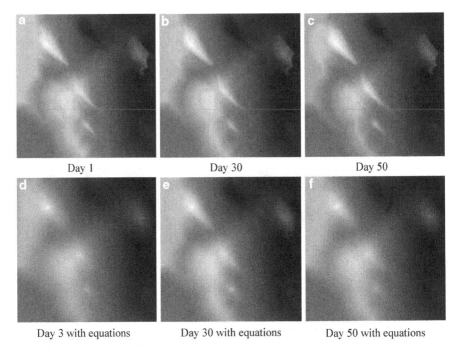

Day 1   Day 30   Day 50

Day 3 with equations   Day 30 with equations   Day 50 with equations

Fig. 10.5 VA groundwater distribution calculated by gradually varied surfaces (**a**)–(**c**) and flow equations (**d**)–(**f**)

References

1. Bouwer H (1978) Groundwater hydrology. McGraw Hill, New York
2. Chen L (1990) The necessary and sufficient condition and the efficient algorithms for gradually varied fill. Chin Sci Bull 35:10
3. Chen L (2004) Discrete surfaces and manifolds: a theory of digital-discrete geometry and topology, Scientific and Practical Computing, Rockville
4. Chen L (2009) Gradual variation analysis for groundwater flow of DC, DC water resources research institute final report. http://arxiv.org/abs/1001.3190
5. Chen L (2010) A digital-discrete method for smooth-continuous data reconstruction. J Wash Acad Sci 96(2):47–65
6. Chen L, Chen X (2011) Solving groundwater flow equations using gradually varied functions (Submitted for publication). http://arxiv.org/ftp/arxiv/papers/1210/1210.4213.pdf
7. Fausett L (2003) Numerical methods: algorithms and applications. Prentice Hall, Upper Saddle River
8. Haitjema HM (1995) Analytic Element modeling of ground water flow. Academic, San Diego
9. Haitjema HM, Kelson VA, de Lange W (2001) Selecting MODFLOW cell sizes for accurate flow fields. Ground Water 39(6):931–938
10. Hamming RW (1973) Numerical methods for scientists and engineers. Dover, New York (Ch2)
11. Harbaugh AW, Banta ER, Hill MC, McDonald MG (2000) MODFLOW-2000, the U.S. Geological Survey modular ground-water model – user guide to modularization concepts and the ground-water flow process: U.S. Geological survey open-file report 00–92, 121 p 2000. http://water.usgs.gov/nrp/gwsoftware/modflow2000/modflow2000.html

12. Johnson C (1987) Numerical solution of partial differential equations by the finite element method. Studentlitteratur, Lund
13. Press WH, Teukolsky SA, Vetterling WT, Flannery BP (2007) Numerical recipes: the art of scientific computing, 3rd edn. Cambridge University Press, New York
14. Pruist GW, Gilding BH, Peters MJ (1993) A comparison of different numerical methods for solving the forward problem in EEG and MEG. Physiol Meas 14:A1–A9
15. Sanford W (2002) Recharge and groundwater models: an overview. Hydrogeol J 10:110–120
16. Stoer J, Bulirsch R (2002) Introduction to numerical analysis, 3rd edn. Springer, New York
17. Yan S, Minsker B (2006) Optimal groundwater remediation design using an adaptive Neural Network Genetic Algorithm. Water Resour Res 42(5) W05407, doi:10.1029/2005WR004303

Chapter 11
Gradually Varied Functions for Advanced Computational Methods

Abstract Gradually varied functions or smooth gradually varied functions were developed for the data reconstruction of randomly arranged data points, usually referred to as scattered points or cloud points in modern information technology. Gradually varied functions have shown advantages when dealing with real world problems. However, the method is still new and not as sophisticated as more classic methods such as the B-spline and finite elements method. The digital-discrete method has another advantage that is to collaborate with these existing methods to make an even better combined approach in applications. We investigate how the gradually varied function can be applied to more advanced computational methods. We first discuss harmonic analysis, B-spline, and finite element methods. Then we give a new consideration for the smooth function definitions in real world problems. This chapter makes more connections to the mathematical aspects of digital functions.

11.1 Similarities Between Gradually Varied Functions and Harmonic Functions

Any constructive continuous function must have a gradually varied approximation in compact space [3, 6]. However, the refinement of the domain for ε-net might be very small. Keeping the original discretization (squares or triangle shapes), can we find some interesting properties related to gradual variation? In this section, we try to prove that many harmonic functions are gradually varied or near gradually varied, This means that the value of the center point differs from that of its neighbor by at most 2. It is obvious that most of the gradually varied functions are not harmonic. We here discusses some of the basic harmonic functions in relation to gradually varied functions.

11.1.1 Review of Concepts

The compatibility between gradually varied functions (GVF) and harmonic functions is important to the applications of gradually varied functions in real world engineering problems.

Recall the concepts we introduced in Chaps. 3 and 9. Let $A_1, A_2, \ldots, A_n$ be rational numbers where $A_1 < A_2 < \ldots < A_n$. Let D be a graph. $f : D \to \{A_1, \ldots, A_n\}$ is said to be gradually varied if for any adjacent pair p, q in D and $f(p) = A_i$, we have $f(q) = A_{i-1}, A_i$ or A_{i+1}. We let $A_i = i$ in this section.

Extending the concept of gradual variation to the function in continuous space: $f : D \to R$ is gradually varied if $|p - q| \le 1$ and $|f_q - f_p| \le 1$, or

$$|f_q - f_p| \le |p - q|. \tag{11.1}$$

On the other hand, a harmonic function satisfies:

$$\frac{\partial^2 f}{\partial x^2} + \frac{\partial^2 f}{\partial y^2} = 0 \tag{11.2}$$

A main property of the harmonic function is that for a point p, $f(p)$ equals the average value of all surrounding points of p.

If f is harmonic, p, q are two points such that $f(p) < f(q)$, and s is a path (curve) from p to q, then we know

$$f_q - f_p = \int_{p,q} \nabla f \cdot d\mathbf{s}. \tag{11.3}$$

If s is a projection of a geodesic curve on f, then does the gradient ∇f maintain some of its properties? For example, is the gradient a constant or does it have any properties relating to gradual variation?

What we would like to prove is that if we define

$$f_{mg}(p,q) = \max\{|\nabla(f)|\} \text{on curve } s \text{ or entire } D, \tag{11.4}$$

would we have

Observation A: $f_{mg}(p,q) < 2 \cdot |(f_q - f_p)| / length(s)$ when f is harmonic?

Therefore, our purpose is to show that many basic harmonic solutions are at least "near" GVF solutions.

11.1.2 Harmonic Functions with Gradual Variation

Given the function value of a set of points J, $J \subset D$ for four-adjacency in 2D (Chap. 2), $f : J \to R$. Using an interpolating process, we can obtain a GVF solution (see Chap. 3) [3]. We can also solve a linear equation using a fast algorithm for a sparse matrix of the harmonic equation based on

$$f_{i,j} = \frac{1}{4}(f_{i-1,j} + f_{i+1,j} + f_{i,j-1} + f_{i,j+1}) \tag{11.5}$$

or give an initial value for f and then do an iteration. This formula gives a fast solution and a definition for discrete harmonic functions [16].

How we use the GVF algorithm to guarantee a near harmonic solution is a problem. We can use the divide-and-conquer method to have an $O(nlogn)$ time algorithm and then iterate it a few times to get a harmonic solution.

Assume b_1 and b_2 are two points on boundary J. $f(b_1) < f(b_2)$ and $s(b_1,b_2)$ is a path from b_1 to b_2. So

$$\frac{(f(b_2) - f(b_1))}{length(s(b_1,b_2))} \tag{11.6}$$

is the average slope of the curve. We can define

$$slope(b_1,b_2) = \max \frac{(f(b_2) - f(b_1))}{length(s(b_1,b_2))}. \tag{11.7}$$

Therefore, the length of $s(b_1,b_2)$ will be the minimum slope. Such a path is a geodesic curve.

With respect to the maximum "slope," the reason for *Observation A* is

$$|\nabla f| \le \left(\frac{\partial f}{\partial x}^2 + \frac{\partial f}{\partial y}^2\right)^{1/2} \le ?2 \cdot slope(b1,b2) \le 2. \tag{11.8}$$

In general,

$$|\nabla f| \le \left(\frac{\partial f}{\partial x}^k + \frac{\partial f}{\partial y}^k\right)^{1/k} \le ?2 \cdot slope(b1,b2) \le 2. \tag{11.9}$$

where $k > 0$. Since $slope \le 1$ is based on the gradual variation condition, we want to show that the harmonic solution is nearly gradually varied. Note that the gradual variation condition is similar to the Lipschitz condition.

There are two reasons for using "2" in the above formula for the ratio: (1) It is not possible for us to use "1," and (2) anything less than 2 is almost gradual variation.

Lemma 11.1. *There are simple cases (under the gradually varied condition) where the harmonic solution reaches a difference of 1.5 comparing to the value of its neighbors.*

Proof. Assume that we have five points in grid space in direct adjacency: $(i,j),(i-1,j),(i+1,j),(i,j-1)$, and $(i,j+1)$, where $f_{i-1,j} = 1$ and $f_{i+1,j} = f_{i,j-1} = f_{i,j+1} = 3$

We want to know what $f(i,j)$ equals. Using the GVF, we get $f(i,j) = 2$ in Fig. 11.1.

Using harmonic functions, we will have $f(i,j) = 2.5$ by (11.5). In the same way, we can let $f_{i-1,j} = 3$ and $f_{i+1,j} = f_{i,j-1} = f_{i,j+1} = 1$. So $f(i,j) = 1.5$ for the harmonic solution, and $f(i,j) = 2$ for the gradually varied solution. $\square$

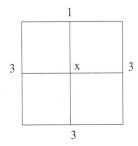

Fig. 11.1 Discrete harmonic interpolation

When we use the harmonic solution to approximate gradual variation, we need to see if we can find the best value when choosing from two possible values. A simple algorithm may be needed to make this decision.

Observation B: There is a GVF that is almost harmonic:
$|center - average of neighbor| < 1$ or $|center - average of neighbor| < c$, where c is a constant.

The above examples show that a perfect GVF is not possible for a harmonic solution. The gradient (maximum directional derivative) is less than $2 \cdot f'_m$, where f'_m denotes the maximum average change (slope) of any path between two points on the boundary that possess the mean of gradual variation.

Every linear function is harmonic. For quadratic functions, we have

$$f(x,y) = ax^2 + by^2 + cxy$$

where the function is harmonic if and only if $a = -b$. However, the following example will not follow this case.

Example 11.1. Three vertices of a triangle are $p1 = (0,0)$, $p2 = (9,0)$, and $p3 = (-8,4)$. The linear function is $f(x,y) = x + 3y$.

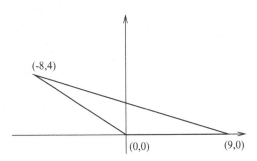

Fig. 11.2 An example of linear functions

This triangle satisfies the gradually varied conditions:

$$|f(p1) - f(p2)| = 9 \le |p1 - p2| = 9$$

$$|f(p2) - f(p3)| = |9 - 4| \le |p1 - p2|$$

$$|f(p1) - f(p3)| = |0 - 4| \le \sqrt{8^2 + 4^2}$$

If we consider a point $p = (x,y)$ on the line $< p_2, p_3 >$ when $x = 0$ and $y = 36/17 = 2.4$, then $f(x,y) > 7$. This point and p_1 do not maintain the condition of gradual variation since $|f(p) - f(p_1)| > 7 > |p - p_1|$.

This example seems to break the observation we made above. However, let us revisit the function $f(x,y) = x + 3y$, letting $z = f(x,y)$. We have $z - x - 3y = 0$. We can have $y = \frac{1}{3}z - \frac{1}{3}x$ represent the triangle and associated function. In general, for a linear function in 3D,

$$ax + by + cz + d = 0.$$

We can always find a coefficient that has the maximum absolute value. So there is an equivalent equation that has

$$AX + BY + D = Z \tag{11.10}$$

where $|A|$ and $|B| \le 1$. This property is often used in computer graphics.

Lemma 11.2. *The Piece-wise linear function preserves the property of gradual variation.*

Proof. We first want to discuss the case of a single triangle where any piecewise linear function is a harmonic function. In this case we can write the function as follows:

$$f(x,y) = ax + by + c, \quad |a|, |b| \le 1$$

So, $\frac{\partial f}{\partial x} = f_x = a$ and $\frac{\partial f}{\partial y} = f_y = b$. The gradient $\sqrt{a^2 + b^2}$ is a constant. There is a horizontal and vertical line that goes through the boundary points. The maximum average rate of change r (the average slope on the path between two points on the boundary) is greater than or equal to $\max a, b$.

Since $|a|, |b| \le 1$, we have $r \le \sqrt{a^2 + b^2} \le \sqrt{2} \max a, b \le \sqrt{2}$. Therefore, $r < 2$.

If this piecewise linear function is on a polygon (2D), then it would still have this property.

The problem is that in this proof, we have not used the conditions of gradual variation directly. The conditions are:

$$|f(p_1) - f(p_2)| = |a(x1 - x2) + b(y1 - y2)| \le |p_1 - p_2| = \sqrt{(x1 - x2)^2 + (y1 - y2)^2}$$

$$|f(p_1) - f(p_3)| = |a(x1 - x3) + b(y1 - y3)| \le |p_1 - p_3| = \sqrt{(x1 - x3)^2 + (y1 - y3)^2}$$

$$|f(p_2) - f(p_3)| = |a(x2 - x3) + b(y2 - y3)| \le |p_3 - p_2| = \sqrt{(x2 - x3)^2 + (y2 - y3)^2}$$

We have used the GVF general property and the triangle constraint. The next section discusses a more general case. □

11.1.3 Gradually Varied Semi-preserving Functions

Now, we can extend the content of the above sections by using more rigorous mathematical definitions. Harmonic functions can be characterized by the mean value theorem. Here we are interested in harmonic functions that are gradually varied. More specifically, a function is said to be gradually varied semi-preserving if

$$\max_D |\nabla u| \le c \cdot \max_{p,q \in \partial D} \frac{|u(p) - u(q)|}{|p - q|} \tag{11.11}$$

where ∇u is the gradient of u, D is a domain with the boundary ∂D, and c is a constant.

The above formula possesses a property of computational importance. We can show that linear functions and quadratic hyperbolic functions satisfy the condition of gradually varied semi-preserving.

If u is linear we can assume $u = ax + by + c$ and if u is quadratic hyperbolic we can let $u = a(x^2 - y^2)$. We do not restrict the values of a, b, and c here.

Lemma 11.3. *If u is linear or quadratic hyperbolic, then*

$$\max_B |\nabla u| \le \sqrt{(2)} \cdot \max_{p,q \in \partial B} \frac{|u(p) - u(q)|}{|p - q|} \tag{11.12}$$

where B can be any circular neighborhood.

Proof. Let u be a linear function $u = ax + by + c$, then

$$|\nabla u| = \sqrt{(a^2 + b^2)} \le \sqrt{(2)} \max\{|a|, |b|\}. \tag{11.13}$$

On the other hand, if we choose $p = (-r, 0), q = (r, 0)$ on ∂B, where r is the radius of circle B, then

$$\frac{|u(p) - u(q)|}{|p - q|} = \frac{|-ar - ar|}{2r} = |a|. \tag{11.14}$$

Choosing another pair p and q on ∂B,

$$p = (0, r), q = (0, -r),$$

we have

$$\frac{|u(p) - u(q)|}{|p - q|} = |b| \tag{11.15}$$

Combining (11.13)–(11.15), we can conclude (11.12) when u is linear.

Now, consider u as a quadratic hyperbolic function: $u = a(x^2 - y^2)$. Then,

$$|\nabla u| = 2|a|\sqrt{(x^2 + y^2)} \le 2|a|r. \tag{11.16}$$

On the other hand, if we choose p and q on ∂B and

$$p = (0, r), q = (r, 0)$$

Then

$$\frac{|u(p) - u(q)|}{|p - q|} = \frac{|-ar^2 - ar^2|}{\sqrt{(2r^2)}} = 2|a|r \tag{11.17}$$

Combining (11.16) and (11.17) we have

$$|\nabla u| \le \sqrt{(2)}\frac{|u(p) - u(q)|}{|p - q|} \le \sqrt{(2)} \max_{x,y \in \partial B} \frac{|u(x) - u(y)|}{|x - y|}$$

and (11.12) follows.   □

Recent studies show an increased interest in connecting discrete mathematics with continuous mathematics, especially in geometric problems. For instance, the variational principle has been used for triangulated surfaces in discrete differential geometry, see [17]. This section presents the idea of combining a type of discrete function (the gradually varied function) with a type of continuous function (the harmonic function) in a relatively deep way. The harmonic function is a weak solution to the Dirichlet problem, which is about how to find a surface from the given boundary curve.

The gradually varied function was proposed to solve a filling problem in computer vision. We are hesitant to use the method of the Dirichlet problem for the discrete filling problem since we do not know the exact formula (function) on the boundary, even though we know the sample points. See Chap. 9 for more information.

11.2 Gradually Varied Functions for Advanced Numerical Methods

Gradually varied functions, as a direct method, is similar to the finite difference method in that they both try to find solutions directly. Other methods, such as the B-spline method and the finite element method, use a basis function to support the reconstruction process by finding the coefficients of a model–A function.

Probably the most remarkable feature of the finite elements method is that it handles complex domain geometries, while the finite difference method is only for rectangular cell decomposition of the domain. For 2D surface reconstruction, the spline method uses tensor products that can only handle rectangular cases as well.

So why is the finite difference method still the most common in industry? The answer is that it is easy in implementation. In addition, it is very efficient in practice. Most reviewers are hesitant to directly point out this fact over the finite elements method. We provide a short review of the finite elements method in this chapter before we attempt to find its connection with the digital-discrete method.

Gradually varied functions can also work with arbitrary domains and any type of decomposition of domains. Looking at this point, the finite difference method is similar to the finite elements method.

11.2.1 The Relationship Between Gradually Varied Functions and the Finite Elements Method

The finite element method has shown great value in the last 50 years in solving differential or integral equations [8, 19, 30]. Compared to the finite difference method, the finite element method is much more narrow. However, this method is more advantageous when the domain comes to the shape not limit to rectangular domains. It can be used to solve any type of domain decomposition problems such as triangulations. The finite elements method needs only a defined boundary in order to solve an equation. Let us first review this method. Consider the Poisson equation,

$$\Delta u = \nabla^2 u = f \tag{11.18}$$

for any smooth function v, we have

$$v \cdot \Delta u = v \cdot f. \tag{11.19}$$

This can be viewed as a type of weak forms of Poisson equation (11.18). The finite elements method is based on Green's (first) identity (also called the divergence theorem),

$$\int_S (\nabla \cdot F) dV = \int_{\partial S} F \cdot da. \tag{11.20}$$

This theorem states that the "volume" of a region can be determined by its boundary. In 2D, S is a region and ∂S is a closed boundary curve. This theorem is a generalization of the fundamental theorem of calculus presented in Chap. 2. The more general version of this theorem is called Stokes' theorem in differential geometry [14].

The finite element method is for any manifold shape, especially for triangulated space when a boundary condition is defined. The idea of the finite elements method is to use simple functions as bases (similar to B-spline) and then represent the solution function using the bases.

Three sample points determine a region with the three sides of a triangle, the Poisson problem can then be solved on this element (the region). The bases function has another use in that it can separate the second order derivative from the first order derivative in an integral equation. The detailed technique is as follows: Let $F = v \cdot \nabla u$. According to Green's Identity

$$\int_S (\nabla \cdot F) dV = \int_{\partial S} F \cdot da.$$

If we select v as zero for all points in ∂S, then $\int_{\partial S} F \cdot da = \int_{\partial S} v \cdot \nabla u da = 0$. Therefore,

$$\int_S (\nabla \cdot F) dV = 0.$$

Thus,

$$\int_S (\nabla (v \cdot \nabla u)) ds = 0$$

and we have $\int_S (\nabla (v \cdot \nabla u)) ds = \int_S (\nabla v \cdot \nabla u) + v \nabla (\nabla u)) ds = 0$. ($(ab)' = a'b + ab'$.) So,

$$\int_S v \Delta u ds = - \int_S \nabla u \cdot \nabla v ds$$

and by (11.19),

$$\int_S v f ds = - \int_S \nabla u \cdot \nabla v ds. \tag{11.21}$$

This formula represents the principle of the finite elements method. (The formula (11.21) is derived from (11.19), so (11.21) is also called the weak form of (11.18).) Let $v = \phi_i$ be a smooth function surrounding the vertex x_i in a (triangulated) decomposition. All triangles that have x_i as the vertex are in $U(x_i)$, an "umbrella" centered at x_i. ϕ_i has the following properties:

(a) $\phi_i = 1$ at vertex x_i;
(b) $\phi_i = 0$ at another vertex x_j, where $j \neq i$ and

ϕ_i is usually linear on each triangle in $U(x_i)$.
If we have n sample points, $< x_i, f_i >, i = 1, \cdots, n$, we can say $u(x) = \sum_{i=1}^n u_i \phi_i(x)$ and $f(x) = \sum_{i=1}^n f_i \phi_i(x)$. By Eq. (11.21), we have,

$$\int_S v \sum_{i=1}^n f_i \phi_i(x) ds = - \int_S \nabla \sum_{i=1}^n u_i \phi_i(x) \cdot \nabla v ds \tag{11.22}$$

Let $v(x) = \phi_j(x)$ for each $j = 1, \cdots, n$, so we have n equations:

$$\sum_{i=1}^n f_i \int_S \phi_j(x) \phi_i(x) ds = - \sum_{i=1}^n u_i \int_S \nabla \phi_i(x) \cdot \nabla \phi_j(x) ds \tag{11.23}$$

We can solve u_j, $j = 1, \cdots, n$. This is a very brief description of the finite elements method. In practice, we need to define the boundary conditions including the standards and first derivatives on the boundary. For a closed manifold, there are no extra considerations. We can solve for u_i and for the function on each sample point when we know each f_i. The Finite Elements principle can also be derived by the variational method [8], minimization of energy functional, based on Spline functions on meshes.

If the 2D domain is not closed, then there is a boundary. The base function v can only define an umbrella $U(x_i)$. Since there is a boundary. v cannot be defined at x_i if x_i is on the boundary. The boundary must be known for f and f', called the function value (the Dirichlet boundary condition) and derivatives (the Neumann boundary condition), respectively.

These are very difficult to know in real world problems. How can it be achieved? Sometimes, researchers need to make assumptions on the boundary. Gradually varied functions can help fill the boundary:

(i) For a value f_i, which is smooth on a manifold, find the convex hull of J and do a GVF fitting. Then, use the FEM.
   We have two more strategies: (a) Based on the convex hull, find the nearest center sample and then do a contour map. For each contour map, use the FEM and then extend the results to the entire D. (b) Use each sample as a center and do a contour map. Based on the contour map, do a FEM, then extend the radius of the neighborhood to cover the entire D. This method would require some average processes since two or more neighborhoods may cover the sample area of the domain.

(ii) If the sample points are too sparse when u_i is obtained, we can use the GVF on u_i's to get the dense smooth function (in a refined way) and then use the finite elements method again. We could also use the GVF on some of f_i to get more values for f_i and then expand the area step by step before using the FEM. The boundary can gradually increase in size to iterate the process.

11.2.2 The Relationship Between Gradually Varied Functions and B-Splines

In Chap. 7, we used the Bernstein polynomial to predict global derivatives. (In fact, we have used the splines in the implementation for real data processing.) We can extend this idea to general splines or B-splines [19]. When a $GVF(f_J)$ is done, then we can perform a B-spline on each horizontal or vertical line. The cubic B-spline ($d = 3$ in (6.18) and (6.19)) provides a good estimation for derivatives up to the third order.

Since $GVF(f_J)$ on a 2D plane is on grid points, we can select the uniform cubic B-spline (meaning that the knots are equidistant) and the formula will be simplified.

Let $A = (f_1, \cdots, f_n)$ be a row in $GVF(f_J)$'s output array. The cubic B-spline formula is:

$$\mathbf{P}_i(t) = \frac{1}{6}\left(1 \ t \ t^2 \ t^3\right) \begin{pmatrix} 1 & 4 & 1 & 0 \\ -3 & 0 & 3 & 0 \\ 3 & -6 & 3 & 0 \\ -1 & 3 & -3 & 1 \end{pmatrix} \begin{pmatrix} f_{i-1} \\ f_i \\ f_{i+1} \\ f_{i+2} \end{pmatrix} \tag{11.24}$$

for $t \in [0,1]$, and $i = 2, \cdots, n-2$. t can either represent the x-axis or y-axis. After the B-spline fitting, $BS(GVF(f_J))$, we can use the results as the solution for the fitting. Another option is that we can continue to calculate the derivatives, such as f_x, f_y, f_xx, f_xy, or f_yy, and then use the method in Sect. 7.4 to get smooth gradually varied surfaces.

The problem is that the curve fitting results the x-axis and y-axis may contain big differences. We can use the average results to reduce error or we can perform a 2D B-spline as discussed in Chap. 6 to find better derivatives.

A problem that must be considered is the resolution of GVF, which should not be too high compared to the final resulting B-spline display. Too many controls or knots will provide less global information for B-spline to calculate. One half or one fourth of the final resolution should be used. For instance, if we want a final 512×512 point array in display, the first GVF fitting should be around 124×124 to 256×256.

11.2.3 Are There Least Square Smooth Gradually Varied Functions?

Data sample points are selected with noise. Therefore, how do we find a fit using gradually varied functions?

A suggested way of doing this is to use gradually varied functions to determine a sample that may be eliminated from the fitting since its difference from the average value is too large and it is not gradually varied. Other research ideas should be encouraged.

Are There Least Square Smooth Gradually Varied Functions? We will also discuss a similar problem in Chap. 12.

11.3 Computational Meanings of Smoothness and Smooth Extensions

A mathematical smooth function means that the function has continuous derivatives to a certain degree $C^{(k)}$. We call this a k-smooth function or a smooth function if k can grow infinitely. Based on quantum physics, there is no such smooth surface in the real world on a very small scale (since there is a distance or gap between two molecules). However, we do have a concept of smooth surfaces in practice since we always compare whether one surface is smoother than another.

This section deals with the possible definitions for "natural" smoothness and their relationships to the original mathematical definition of smooth functions. The motivation behind giving a definition for a smooth function is to study smooth extensions in practical applications.

We observe this problem from two directions: (1) From discrete to continuous: We suggest considering both micro smooth, the refinement of a smoothed function; and macro smooth, the best approximation of a smoothed function using existing discrete space. For two or higher dimensional cases, we can use Hessian matrices when use Taylor expansion. (2) From continuous to discrete: We suggest a new definition for being naturally smooth, which scans from down scaling to up scaling in order to obtain the ratio for sign changes (ignoring 0) to represent the smoothness. For differentiable functions, mathematical smoothness does not necessarily mean " looking" smooth for a sampled set in discrete space.

Finally, we discuss Lipschitz continuity for defining smoothness, which is called discrete smoothness. This book gives philosophical consideration for smoothness in practical problems rather than a mathematical deduction or reduction, even though our inferences are based on solid mathematics.

11.3.1 Overview of Continuity and Discontinuity

The world can be described by continuity and discontinuity. While most parts are continuous, discontinuity can provide the boundary for continuous components. Likewise, continuity may give context for discontinuity. For instance, an image can be separated into several continuous components. The edge of a component indicates the discontinuity of two parties. However, the boundary itself could be continuous (continuously colored).

Continuity sometimes means controllable, while discontinuity may mean uncontrollable, such as something suddenly happening. For example, a sunny day suddenly becomes rainy.

When continuity is not good enough to describe a phenomenon, we need to define smoothness. This is the same thing in discontinuity, where we want to define chaos [11, 15, 24].

Mathematicians and scientists always try to make things simpler. What exactly is smoothness? A mathematical smooth function means that a function has continuous derivatives up to a certain degree $C^{(k)}$. In Chap. 2, we call it a k-smooth function. A function is called a smooth function if k can be infinite.

Based on quantum physics, there is no perfectly smooth surface. That means such a smooth surface does not exist in the real world on the quantum-scale. So the "natural" smoothness is relative and not absolute.

On the other hand, a mathematical smooth function does not always mean continuity in the real world. For instance, when a sunny day turns to rain, the rain does not usually come down consistently at 1 cm a minute. The rain usually has a few drops and then becomes heavier. Under close observation, the weather can be said to change continuously or smoothly. However, in reality, the sunny day suddenly became rainy, which is discontinuous. Therefore, "continuity" depends on the scale and sampling.

The problem is, how can we define smoothness that cannot only be used in real world problems but also does not cause a contradiction mathematically? In fact, mathematicians have already observed this problem. They used Lipschitz conditions (see Chap. 2) and rectifiability to restrict a function's continuity and smoothness [9, 23].

In terms of Lipschitz extension, see Chap. 7, it was amazing that three papers were published in the same year dealing with the same or very similar problems, as McShane, Whitney, and Kirszbraun all did independently in 1934. They discovered an important theorem for the Lipschitz function extension. What they proved is: For any subset E of a metric space S, if there is a function f on E that satisfies the Lipschitz condition, then this function can be extended to S while preserving the same Lipschitz constant [12, 18, 27, 28].

So is our problem solved? Actually it is not. The McShane-Whitney or McShane-Whitney-Kirszbraun theorem only provides a theoretical result. This is because the Lipschitz constant could be very big. No one involved in real world problems would want to admit that such a function is "continuous."

In 1983, Krantz published a paper that specifically studied smoothness of functions in Lipschitz space for technical applications [13]. Unfortunately, such an important movement did not get enough attention in mathematics.

On the other hand, along with the fast development of computer science, scientists started to study "continuous functions" in a digital format for image processing. Rosenfeld pioneered this direction of research [20, 21]. In 1984, he used a fuzzy logic concept to study the relationship among pixels [29]. He defined the concept of fuzzy connected components. In 1985, Chen proposed another type of connectedness between two pixels called λ-connectedness (see Chap. 12) [1, 4]. This concept was later found to have close connections to continuous functions. In 1986, Rosenfeld proposed discretely continuous functions [22]. In 1989, Chen simplified λ-connectedness to gradual variation and proved the necessary and sufficient condition of the extension [2].

Chen's constructive method suggests an algorithm that is totally different from McShane's construction, which builds a function from maximum or minimum possibilities. The algorithm designed by Chen uses local and middle construction, which is more suitable in practice. In addition, Chen used a sequence of real or rational numbers instead of integers, treating the problem more generally. As we have discussed in Chaps. 7 and 8, this method overcomes the limitations of Lipschitz space by enlarging the space to local Lipschitz [2, 3, 5, 6, 13].

Even though we have a continuous or mathematical smooth function to describe the addressed question, the problem may only be able to be presented in a finite form for a large number of applications using finite samples in today's technology.

Based on the McShane-Whitney-Kirszbraun theorem, for a bounded space, a "continuous" surface always exists. According to the classic Weierstrass approximation theorem: If f is a continuous real-valued function on $[a, b]$, a closed and bounded interval, then f can be uniformly approximated on that interval by polynomials to any degree of accuracy. In addition to the results obtained by Fefferman [10], differentiable extensions of Whitney's problem do exist. However, the solutions are based on infinite refinement of the domain space. In practice, we have limited availability for the space we are dealing with. Recall the example of the sunny day that turns to rain. Even though we could insert a variable time, accurate to the microsecond, we still have a discontinuous event that occurs.

The mathematical definition of smooth functions does not provide a precise description for such a phenomenon in real problems. The existence of a smooth function for "sudden rain" is not good enough for the problem if our sampling scale is in minutes or seconds. Smoothness is dependent on sampling in the real world and its applications. In many cases, absolute smoothness based on the mathematical definition may not work correctly.

We propose possible definitions for natural smoothness and their relationship with the original mathematical definition of smooth functions. The motivation of giving the definitions for smooth functions is to study smooth extensions in practical applications.

11.3.2 Micro- and Macro-smoothness: From Discrete Functions to Continuous Functions

In industry, there some considerations already exist, which are called micro-smoothness and macro-smoothness. They are used for measuring similar things in real world problems. In the paper and fabric industry, Dimmick described that *macro-smoothness* is based on base paper deformation while *micro-smoothness* is based on coating [7]. Micro-smoothness is a local smoothness, and macro-smoothness is a global smoothness. Micro is coated smooth on the surface while macro is materially and intrinsically smooth. Micro is externally smooth and macro is internally smooth [7, 25].

A somewhat general form of the question is as follows: Let us assume a discrete function $f : D_0 \to R$. How do we make $F : D \to R$ smoother than f where $D_0 \subset D$ and $|D| < c|D_0|$, where $c > 1$ is a constant?

11.3.2.1 Micro-smoothness

If we try to get a micro smooth function, we can first use piecewise linear functions to interpolate the data and then use the Bernstein polynomial to fit the data as described in Chaps. 2 and 6. We could also use the method in [10]. The method Fefferman adopted uses Whitney's jets at each point, then uses a binary tree structure to combine them. Each step of the union operation is to generate a system of linear inequalities based on the Lipschitz conditions for every degree of derivative. A Whitney's jet is a local expansion of the Taylor series on a refined grid or mesh.

11.3.2.2 Macro-smoothness

The macro smooth function is to keep (or make small changes to) the original grid or meshes. If we assume the original function is Lipschitz with Lipschitz constant *Lip*, then we have to make changes to get a smooth function in terms of using finite difference instead of derivatives. In other words, using approximation (with limited errors) instead of equal left and right derivatives.

Here is something to consider: when we make a macro smooth function, there is no need to change values for every point; we can select some points that will not be changed (similar to guiding points in application). In such a case, it is equivalent to the extension of finite sampling cases. We used this methodology for algorithm design in [13].

11.3.2.3 Relationship with Existing Concepts

In the discussion of natural smoothness of continuous functions, micro smoothness is "smoothness" in high resolution (dense sampling) and macro smoothness is smoothness in low resolution. One could say these are the high frequency and low frequency components in Fourier and wavelet analysis. What is the difference?

We are not very interested in every single frequency, similar to Fourier and wavelet analysis. We are more interested in the extreme cases such as the maximum and minimum (possible "frequencies") that have micro and macro "smoothness." Where to find these points is key. We could use persistent analysis to determine the values if necessary. Thus,

$$Function = \text{micro component} + \text{macro component} \qquad (11.25)$$

This is an approximation to the real data.

11.3.3 Natural Smoothness of Continuous Functions: From Continuous Functions to Discrete Functions

Since a continuous function has a uniform approximation from a sequence of polynomial functions, the mathematical smoothness of a continuous function does not really mean much. From continuous to discrete, we suggest a new definition for natural smoothness that includes scanning from large scaling to small scaling in order to get the ratio for the sign changes of the derivatives on the curve (ignoring zero) to represent smoothness. For differentiable functions, mathematical smoothness does not necessarily mean "looking" smooth for a sampled set in discrete space. (If the function is a rectifiable curve, such a sampling always exists [9]).

A function that is not continuous could be a Lipschitz function that has a small Lipschitz constant. It may have good natural smoothness. In other words, after sampling, the macro smoothness is good, but the micro smoothness may not be good. This is exactly like material smoothness (such as for paper and cloth) and its relation to deformable properties.

Another good example is that a mountain looks smooth on a skyline, but when we get close to the mountain, we see a lot of trees and that it is not smooth at all. So the natural smoothness is relative to resolutions and scales. For a one-dimensional function, assume that:

Nsamples: denotes the number of samples selected in the function.

DerivativeSignChanges: denotes the total number of sign changes in the derivative function of the reconstructed function based on the samples by the finite difference method, McShane-Whitney mid function (see Chap. 7), or gradually varied functions.

Definition 11.1. Natural smoothness of a continuous 1D function is a (stable) ratio:

$$R = (Nsamples - DerivativeSignChanges)/Nsamples \qquad (11.26)$$

The worst case scenario of this definition is for the function with the following pattern at its sampling points $\{-1, +1, -1, +1, \ldots\}$. The smoothness is almost 0, as shown by the limits of this formula and the probability of the estimation. Statistical methods such as the student t-test may be needed to find the confidence region. When a small set is sampled, the micro-changes will not affect the value of the natural smoothness. It will overcome the inconsistencies of classical mathematical definitions for smooth functions. Natural smoothness is a concept with relativity.

Let us define the natural smoothness for 2D and high dimensional cases. We still use sampling and reconstruction in order to then find smoothness. There are two steps to obtaining the smoothness: (1) Get sample points, *SN*, and their values. (2) Reconstruct the function using the McShane-Whitney mid function or the gradually varied function. (3) Use Hessian matrices to find the extreme points of the reconstructed functions. Count the number *EN*.

Sampling is necessary since we want to eliminate the high frequency parts based on the average sampling scale. Another method uses traditional wavelets or Fourier transforms to get the decomposition of frequencies, and then do a reconstruction based on different frequency scales. We can still calculate the number of extreme points using Hessian matrices.

Definition 11.2. Natural smoothness of a continuous kD function is a (stable) ratio:

$$R = \frac{SN - EN}{SN} \tag{11.27}$$

11.3.4 Natural Smooth Functions for Discrete Sets and Gradually Varied Smooth Reconstruction

If we have a finite number of samples and no other information, we may only be able to fit a continuous function using gradually varied fitting. If we want the fitting to be good and smooth looking, we could polish the function. Based on the definition of natural smoothness, the linear interpolation (if one dimensional), or gradually varied fitting, and polishing will not change the natural smoothness of the fitted function. They are essentially similar functions.

Using gradually varied functions for smooth reconstruction can either be micro- or macro- smooth reconstruction. Since for most applications we like to keep the original grid and most existing reconstruction used the polynomial method that can yield a mathematically smooth function, we mainly use gradually varied functions for macro smooth functions.

We discussed natural smoothness in Sect. 11.3.3. The natural smoothness could change when the grid scale changes if we use gradually varied functions. What we want to obtain is a fitted function that has the best uniform approximation to the guiding points.

We used this method in Chap. 7. The easiest way is to use the finite difference method to calculate the left and right derivatives for higher order derivatives. This finite difference method will not change the extreme points. Therefore, it would maintain the natural smoothness since natural smoothness is macro smooth in down scaling and is micro smooth in up (dense) scaling.

Since the case is in the discrete set, it is hard to obtain the exact same value from the left and right derivatives. We only want to limit the difference between them for a given error *epsilon*. Related ideas are presented in [10, 13]. We want,

$$\left| \frac{f(x_{i+1}) - f(x_i)}{x_{i+1} - x_i} - \frac{f(x_{i+2}) - f(x_{i+1})}{x_{i+2} - x_{i+1}} \right| \leq epsilon \tag{11.28}$$

If f is Lipschitz, and the x_i are equidistant, we have

$$|2f(x_{i+1}) - f(x_i) - f(x_{i+2})| \leq epsilon \tag{11.29}$$

Krantz showed this estimation using an example in [13] .

In summary, without additional information, such as a fitting followed by a formula or differential equations, the finite difference method for regular grids or gradually varied functions for irregular sampling are the best ways. In addition to the continuous methods, we can use a polishing method to smooth the functions.

11.3.5 Discrete Smoothness, Differentiability and Lipschitz Continuity

Natural smoothness only counts the general number of extreme points. It may not give a detailed description of discrete or digital functions. Based on the property of a polynomial, we know that the $(n+1)$th order derivative of a degree n polynomial will be 0.

A discrete function is always near a polynomial in terms of approximation as we described above. Now, we define a discrete smoothness to simulate the smoothness of polynomials.

Let's consider a function in an equidistant sampling system, or a function $f : \{1, 2, \ldots, n\} \to R$.

$$|f(x) - f(y)| \le Lip|x - y| \tag{11.30}$$

Assume that Lip is the Lipschitz constant (the smallest constant satisfying the above equation). Let $f(x) = f^{(0)}(x)$ and $Lip = Lip^{(0)}$, so we can define the simple difference $f^{(k+1)}(x) = f^{(k)}(x+1) - f^{(k)}(x)$. The Lipschitz constant for $f^{(k+1)}$ is

$$Lip^{(k)} = \max \frac{|f^{(k)}(x) - f^{(k)}(y)|}{|x - y|} \tag{11.31}$$

for all x and y, where x and y are integers or $x = k/m$ and $y = k/m$.

We can show this by assuming $x > y$,

$$
\begin{aligned}
|f^{(k)}(x) - f^{(k)}(y)| &= |f^{(k)}(x) - f^{(k)}(x-1) + f^{(k)}(x-1) - f^{(k)}(x-2) + \\
&\quad \cdots + f^{(k)}(y+1) - f^{(k)}(y)| \\
&\le |x - y| \max |f^{(k)}(u+1) - f^{(k)}(u)|
\end{aligned} \tag{11.32}
$$

where $(y+1) < u < x$. By definition, $f^{(k+1)}(x) = f^{(k)}(x+1) - f^{(k)}(x)$,

$$|f^{(k)}(x) - f^{(k)}(y)| \le |x - y| \max |f^{(k)}(u+1) - f^{(k)}(u)|$$

So, we can prove that

$$Lip^{(k)} = \max |f^{(k+1)}(x)|.$$

This is consistent with the idea that the Lipschitz constant can be viewed as the largest derivative of the function.

What we expect is that there exists a k_0 such that when $k > k_0$, we have

$$Lip^{(k)} > Lip^{(k+1)}.$$

Based on this, we suggest the following definitions:

Definition 11.3 (absolute). A discrete function is called a discrete smooth function if there is a K such that $Lip^{(k)}$ defined in (11.30) is 0 when $k > K$.

Definition 11.4 (almost). A discrete function is called a discrete smooth function if there is a K such that $Lip^{(k)}$ defined in (11.30) is smaller than $c_2/(2^{(k-c_1)})$ when $k > K$, where c_1 and c_2 are constants.

Definition 11.5 (k-discrete smooth function). A discrete function is called a K-th-order discrete smooth function if there is a K such that $Lip^{(k)}$ defined in (11.30) is smaller than $c_2/(2^{(k-c_1)})$ when $k <= K$, where c_1 and c_2 are constants and c_2 is linear to $Lip^{(0)}$.

11.4 Future Remarks

The smooth mapping between two manifolds is at the center of many studies in mathematics, especially in differential geometry and functional analysis. The current tendency shows that discrete methods are significantly involved in related research.

However, more practical or feasible algorithms are still weak in real problems and their applications.

References

1. Chen L (1985) Three-dimensional fuzzy digital topology and its applications(I). Geophys Prospect Pet 24(2):86–89
2. Chen L (1990) The necessary and sufficient condition and the efficient algorithms for gradually varied fill. Chin Sci Bull 35(10):870–873
3. Chen L (1990) Gradually varied surfaces and gradually varied functions, in Chinese. In English 2005 CITR-TR 156, University of Auckland
4. Chen L (1991) The lambda-connected segmentation and the optimal algorithm for split-and-merge segmentation. Chin J Comput 14:321–331
5. Chen L (1992) Random gradually varied surface fitting. Chin Sci Bull 37(16):1325–1329
6. Chen L (2004) Discrete surfaces and manifolds. Scientific and Practical Computing, Rockville
7. DimmiCk AC (2007) Effects of sheet moisture and calendar pressure on PCC and GCC coated papers. Tappi J 6(11):16–22.
8. Fausett L (2003) Numerical methods: algorithms and applications. Prentice Hall, Upper Saddle River
9. Federer H (1969) Geometric measure theory. Springer, Berlin/Heidelberg/New York, p 202
10. Fefferman C (2009) Whitney's extension problems and interpolation of data. Bull Amer Math Soc 46:207–220

11. Gollub JP, Baker GL (1996) Chaotic dynamics. Cambridge University Press, Cambridge/ New York
12. Kirszbraun MD (1934) Über die zusammenziehende und lipschitzsche transformationen. Fund Math (22):77–108
13. Krantz SG (1983) Lipschitz spaces, smoothness of functions, and approximation theory. Expos Math 3:193–260
14. Kreyszig E (1991) Differential geometry. Dover Books on Mathematics, New York
15. Li TY, Yorke JA (1975) Period three implies chaos. Am Math Mon 82:985–92
16. Lovasz L (2004) Discrete analytic functions: a survey. Surveys in differential geometry IX. In: Grigor A, Yau S-T (ed) Eigenvalues of laplacians and other geometric operators. International Press, Somerville
17. Luo F (2008) Variational principles on triangulated surfaces. http://arxiv.org/abs/0803.4232v1
18. McShane EJ (1934) Extension of range of functions. Bull Am Math Soc 40:837–842
19. Quarteroni A (2009) Numberical models for differential problems. Springer, Milan/New York
20. Rosenfeld A (1979) Fuzzy digital topology. Inform Control 40:76–87
21. Rosenfeld A (1984) The fuzzy geometry of image subsets. Pattern Recognit Lett 2:311–317
22. Rosenfeld A (1986) 'Continuous' functions on digital pictures. Pattern Recognit Lett 4:177–184
23. Schwartz JT (1969) Nonlinear functional analysis. Gordon and Breach Science Publishers, New York
24. Sharkovskii AN (1964) Co−existence of cycles of a continuous mapping of the line into itself. Ukr Math J 16:61–71
25. Taylor CJ (2003) Advanced machine clothing to optimize board smoothness and machine efficiencies. In 57th Appita annual conference and exhibition, Melbourne, Australia 5–7 May 2003 Proceedings. Appita Inc., Carlton, pp 17–23
26. Valentine FA (1943) On the extension of a vector function so as to preserve a Lipschitz condition. Bull Am Math Soc 49:100–108
27. Valentine FA (1945) A Lipschitz condition preserving extension for a vector function. Am J Math 67(1):83–93
28. Whitney H (1934) Analytic extensions of functions defined in closed sets. Trans. Am Math Soc 36:63–89
29. Zadeh LA (1965) Fuzzy sets. Inform Control 8:338–353
30. Zienkiewicz OC, Taylor RL (1989) The finite element method. McGraw-Hill, London

Chapter 12
Digital-Discrete Method and Its Relations to Graphics and AI Methods

Abstract As a practical method, digital-discrete data reconstruction uses both discrete and continuous methods for data interpolation and approximation. Today, a significant development in discrete mathematics is digital technology. Digital methods contain a flavor of graphical presentation and artificial intelligence. It is very interesting to explore the relationship between digital methods and graphical and intelligence methods. In this chapter, we introduce the subdivision method and the moving least squares method. These two methods are popular smooth data fitting methods in computer graphics. The subdivision method is an intuitive method for smooth shape design; the moving least squares method is a mesh-free method for data fitting. Our purpose is to provide a potential link from main stream techniques to digital functions. We also present the extension of digital functions to more general cases in artificial intelligence, especially in partial information searches and image segmentation. This expansion uses lambda-connectedness, which shares the same mathematical foundation as gradually varied functions. We include the algorithms for segmentations and fitting for image objects. This chapter also contains future research topics of the digital-discrete method.

12.1 Subdivision Surfaces Versus Gradually Varied Surfaces

Computer graphics is an area that uses computers to design and display geometric shapes or user interfaces [25]. The subdivision method was introduced in 1978 and is related to the B-spline method for geometric design [3, 21].

Geometric design usually has a very different meaning from geometric data reconstruction. Geometric design means that the designer knows what the shape should look like or the shape he is looking for. The goal of reconstruction is to find out what the true shape is.

We are interested in subdivision methods because they are very popular and effective methods in computer graphics. Also, they can help get smooth surfaces based on guiding points.

L.M. Chen, *Digital Functions and Data Reconstruction: Digital-Discrete Methods*,
DOI 10.1007/978-1-4614-5638-4_12, © Springer Science+Business Media, LLC 2013

Subdivision uses the strategy of adding new data points (refinement) on original meshes then using the surrounding points to modify the original points, thereby cutting the corner off. An alternate method, sometimes called the 4-point algorithm, smoothes the corner points (filling the neighbors, an interpolation method). These two methods are essentially the same in terms of having the same mathematical foundation. They will both generate C^1 functions (C^1 on extraordinary points and C^2 on the rest of the function).

Most of the time the modification is a type of weighted average of the related points. This is called a scheme. Different schemes generate different results; in other words, different schemes work for different problems. There are other schemes other than the Catmull-Clark scheme [3] and Doo-Sabin scheme [21]. These schemes are introduced based on certain mathematical principles. Therefore, the subdivision method is a local average method.

The gradually varied surface method is a global method. Using the gradually varied method first and then the subdivision method can be useful in some problems and could be a new direction for future research.

12.1.1 The Principle of the Subdivision Method

Cutting the corner off is the basic idea of the subdivision method. The "cut" generates new corner points (and also more points) that are smoother than the original ones. Recursively applying this technique will result in a smooth curve or surface.

The following example demonstrates this idea: For a square shape, Fig. 12.1a we first cut the four corners off. We get Fig. 12.1b; then, cut the eight corners off and we get Fig. 12.1c. We will eventually get a circle-like shape. There are four points that will not change at each side called extraordinary points (it is usually true that after the first cut, the number of extraordinary points remains the same).

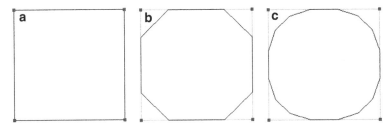

Fig. 12.1 Example of subdivision: (**a**) the original curve, (**b**) cut four corner off, and (**c**) cut eight corner off

This idea was first presented by Chaikin in 1974 [4]. It had some similarity to Bezier curves. Bezier curves are generated by control polygons that use Bernstein polynomials. However, Chaikin used control polygons directly for corner cutting.

For the most popular Catmull-Clark scheme, the formula is the following: All vertices of the mesh refer to original points. For an original point p (with $v(p)$ as the value), we do the following process:

(1) Set new edge points at the mid point of each edge that links to p (using the average value for the function). The average value is denoted by E. (2) Set new face points in the centroid of each face that contains p. The average value is denoted by F. Next, set the values of the original points to the new values v_{new} as

$$v_{new}(p) = \frac{v(p)(n-3)+F+2E}{n} \tag{12.1}$$

where n is the number of faces linking to p. All the original points will be set to the updated points, so the process will run repeatedly until no corner can be cutted. The results obtained will be first order differentiable at extraordinary vertices and second order differentiable at other points.

Corner cutting is an approximation process. The subdivision algorithm can also be modified for interpolation. This can be done by inverting the process by adding two triangles (not cutting a triangle) in each side edge of one dimension. For surfaces, add a 3D simplex at the top of each face to make the corner smoother.

For curve interpolation, this algorithm is called the 4-point algorithm. The algorithm takes two neighbors on each side when inserting a new point. The four old points can make a third degree polynomial . Then, the parametric midpoint is defined as the new point (Fig. 12.2).

Fig. 12.2 Example of four-points interpolation: (**a**) the original curve, (**b**) add control points on each side of corner points, and (**c**) add more control points

12.1.2 Gradually Varied Surfaces and Subdivision Surface Combinations

The subdivision method is a local fitting method. However, the subdivision method can only make C^1 (or C^2) functions. It may not be appropriate to use data reconstruction directly since it may change the location of the original points quite a bit when the sample data is not dense.

Gradually varied functions contribute to subdivision methods by first fitting the gradually varied surfaces. If there are relatively dense points, then continue to make the subdivision surfaces. This is expected to get better results since the gradually varied surface method is a nonlinear method. In other words, gradually varied fittings can help the subdivision when the subdivision method is intended to be a data reconstruction method.

In fact, most uses for the practical methods to make smooth surfaces on a manifold are in computer graphics.

The subdivision method is a refinement method. It usually applies to ordinary shapes such as triangles, circles, and rectangles. Other methods for the same purpose use local rectangular or circular refinements. However, these methods are not true interpolation and extension methods. They are designed for refinements, meaning that we have to have relatively good existing decompositions or meshes before applying these methods.

The idea of subdivision methods can also be used in digital-discrete methods. If multi-scaled or domain refinement is allowed in gradually varied fitting (see Chap. 7 and [7, 11, 12]), we can add two data points to smooth the value at a given point. For example, after the initial fitting $GVF(x)$, if the left derivative and the right derivative at point x_0 are very different, then we can add two data points to cut down the extreme value $GVF(x_0)$ lower or lift the both sides of $(x_0, GVF(x_0))$ to smooth out the value at x_0. If we "cut off the corner," then we will continue on to the gradually varied fitting by establishing gradually varied derivatives like in Chap. 7. "Cutting off the corner" is a fast way for smooth approximation.

12.2 Mesh-Free Methods Versus Domain Decomposition

For randomly arranged sample points on a plane or manifold, we can assume a sample data set $S = \{(p_i, v(p_i)) | i = 1, \cdots, n\}$ where p_i is a point on the manifold and $v(p_i)$ is a real number or vector; S is finite.

There are two ways to solve the problem of data reconstruction: (1) Find the analytical solution, i.e. we find the exact solution if it exists and can be expressed; (2) Find the numerical solution where we can usually only get one approximation. In most cases, real problems are not solvable analytically. As a result, the primary concern today is still the numerical solution.

The methods for this problem can be divided into two categories: mesh-dependent and mesh-independent. The mesh-dependent method requires convex-cell decomposition, triangulation, or gridding on the domain based on the sample points, e.g. the methods introduced in Chap. 6. The mesh-independent method attempts to fit the data without predefined grids or triangles, but it requires a dense sample point set. It has become somewhat popular in recent years especially in computer graphics [1, 26, 27].

12.2.1 The Moving Least Square Method

In mesh-free methods, the moving least square technique is the most adopted. It uses a circular weighted neighborhood instead of triangles [1, 26, 27].

Assume $S = \{(p_i, v(p_i)) | i = 1, \cdots, n\}$ is the set of sample points. For a point p in domain D, select some samples in S that surround p. Find a best fitted polynomial f of degree k based on these sample points. The selection of surrounding points is determined by a Gaussian distribution-like function, for instance $\theta(p - p_j) = e^{-\frac{(p-p_j)^2}{a}}$. This means the contribution (weight) of point p_j tends to 0 as $|p - p_j| \to \infty$.

Formally, we want a polynomial f with degree k that minimizes the weighted least-square error at point p:

$$\sum_{i \in S} (f(p) - p_i)^2 \theta(p - p_i) \tag{12.2}$$

When p moves, we will get the entire function. This fitting must be preformed at each location p so that in practice, we will need to select each point to be fitted. This process is slow since we must fit one point at a time. Since each fitting covers the entire area, we can also select some of the points in the domain to fit the functions, then use a partition of unity to combine the functions at each point in D.

The moving least square (MLS) method contains a good philosophy in dealing with randomly arranged sample points. It is a reasonable and practical method. However it requires a dense sample data set, otherwise there might not be enough points to do a local fitting.

MLS requires the consideration of three questions in application: First, we need a balanced sample point or a relatively even distribution. Consider the Dirichlet problem: given a boundary of sample points, you are asked to fill a surface. How would the weight function be chosen?

Second, even though a Gaussian distribution weighted function was selected in general cases, if the surrounding region considered is too small, we would not have enough guiding points, see Fig. 12.4. This is why sometimes in artificial intelligence, k-nearest neighbors are required, but we would then need to determine the value of k.

Third, because there is no mesh, for a thin manifold, MLS may fail to determine the neighbor points by using Euclidean distance. That is to say, for any fitting problem on a manifold, a system that recodes the adjacencies must be defined before the interpolation or fitting. Being mesh-free is possible, but being adjacency relation free is not possible for such a calculation. This is why we can say that MLS is not a general-purpose interpolation method.

When we are given a set of sample points, do we need to consider an artificial intelligence method when deciding on how to make a reasonable $\theta(p - p_i)$ function?

MLS (somehow likes Voronoi map) is based on the measurement of a predefined weight function. This function considers guiding points in the calculation based on weight that decreases by distance from the point considered. It is a good idea for some applications, but would not work in all cases. The principle is still based on the linearity of domain separability. Even though there is a nonlinear weight function added, philosophically, the method is still "linear" since the importance of the function decreases as distances increases. In fact, MLS cannot automatically choose the weight functions (Fig. 12.3).

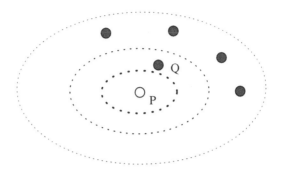

Fig. 12.3 *P* is the moving point (of unknown value). *Dotted ovals* are the highly weighted areas in which the location is contributing heavily to the value of point *P*. *Q* and other *solid dots* indicate the sample sites

In Fig. 12.4, *P* is the moving point. The dotted ovals are the highly weighted areas in which the sites (sample points) are contributing greatly to the value of point *P*. The cross *X* marks the sites.

The value of *P* might only be dependent on *Q*. However the outer four points will contribute evenly to *P*. In such a case, it would be easy to generate false interpolation results.

For some *P* there are only one or two sites in the closed neighborhood. But how do we determine these cases? What type of artificial intelligence or pattern recognition methods should be selected to dynamically determine θ function [34, 35]?

We can see that MLS does not depend on meshes but on circles, so there must be dense samples on the manifolds. Therefore, the existing practical methods for "linear" decomposition or smooth refinements will not be accurate.

12.2.2 The Methods Based on Domain Decomposition

Mesh-free methods have certain advantages over methods based on domain decomposition. We have presented several surface reconstruction approaches in Chaps. 6 and 7.

Why do we need domain decomposition? The reason is that we need to first determine the region affected by a sample point. It has similar effects or functionalities to the weight function in mesh-free methods.

The most natural decompositions include the Voronoi diagram and its dual correspondence, the Delaunay triangulation. Based on a set of sample points in a region, called sites, the Voronoi diagram partitions the domain into sub-regions where each point of the domain is categorized into a subset containing a site. The point is then considered closer to this site than to any other site. Its Delaunay triangulation is used to link sites that have adjacent cells.

If decomposition must be done, then there is no better method other than Voronoi diagram or Delaunay triangulation. These two mechanisms reach the excellence based on human's intuitive judgments. The convex sub region as it is in Voronoi diagram maintains a number of good properties. Many real world examples are available, including the popular nearest neighbor method. Mount gives an excellent explanation of these questions in his lecture notes [29]. Another way of defining a Delaunay triangulation is to do a triangular decomposition on the sites such that no site is inside the circumcircle of any triangle, see Chap. 6.

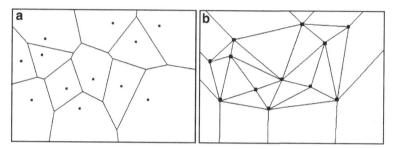

Fig. 12.4 Voronoi and Delaunay diagrams: (**a**) A Voronoi diagram; (**b**) Delaunay triangulation on the same sites

Recently, Fefferman proved the existence of a smooth function up to a certain smoothness degree (order) based on the refinements of grid points in Euclidean space applicable even in extended dimensions [14]. Fefferman et al. described a method that is based on the local Whitney's jet problem and has some similarities with MLS. A Whitney's jet is a polynomial in a tangent space at point p in the domain. Fefferman used the partition of unity for smoothing the "boundary" of each Whitney's jet. These methods mainly use linear methods or are based on linear methods.

For triangulated domains, we can still apply a refinement method. If we know the function values at three vertices, we can insert three more middle points on each edge. Then we can use the mean of the two end points to find the function value of the middle points. Therefore, we have a continuous interpolation. This is very similar to smooth color interpolation in graphics [13, 25].

The advantage of these methods is that they can do localized fittings. On the contrary, the finite difference method can only do global fittings, meaning that every sample point must be considered for each undetermined new data point. Global fittings using the finite element method or moving least squares method are too costly in practice. Another disadvantage of using these methods is that they would not be able to deal with a very small set of sample points.

Even though we can use a gradually varied fitting first to assist MLS by adding more sample points, we are looking for more interesting solutions by proposing local smooth gradually varied jets.

12.2.3 Local Smooth Gradually Varied Jets?

If we have dense sample points, can we build a local gradually varied function centered at a point p before moving point p in the domain to obtain a smooth fitting? This idea comes from the Whitney's jets problem and the moving least squares method.

What is a local smooth gradually varied function? It is a natural spline or B-spline function based on the $GVF(f)$ around p, see Chap. 11. Using the idea of B-splines, make a local smooth gradually varied function for each region by using a linear combination to merge them into the final solution.

The moving least squares method uses local sample points to fit a polynomial with a Gaussian shape or uses the Gaussian shape as weights at the selected point. The idea behind this method has some similarities to the local regression method originally proposed in [20].

Different from using Gaussian distribution-shapes as weights, gradually varied reconstruction is a natural mesh free fitting method. We can also add a Gaussian weight to each of the other points.

12.3 Expanding Gradually Varied Functions to Image Segmentation

Image based decomposition is called image segmentation. It is a fundamental approach in image processing and computer vision [24, 30, 33]. Image segmentation partitions an image into different components called objects. The criteria for the segmentation varies and it depends on the purpose of the image processing as well as the properties of the image. Gradually varied functions can be extended to do image segmentation. This is because finding a similar or gradually varied part of an image can be viewed as tracing an object in an image.

The general form of gradually varied functions is called λ-connectedness. λ-connectedness was originally used as a technique to search layers in three dimensional seismic data [8, 9, 18]. Two geometric points are determined to be in the same layer if there is a path that links them and their intensity values on the path change gradually (λ-connected). Similarly, λ-connectedness was also used to segment an image or to find an object in the image. On the other hand, if we know some sample points in a system with the property of smooth variation (among the individuals in the system), how do we generate a whole system with λ-connectedness?

In the following subsections, we give a quick review of λ-connectedness and present an image segmentation algorithm. Then, we propose a new method called the λ-band-connected search for noisy systems. At the end of the section, we study the λ-connected fitting method with gradients for smoothness.

12.3.1 λ-Connectedness and Classification

λ-connectedness can be defined on a graph $G = (V, E)$ with an associated (potential) function $\rho : V \to R^m$, where R^m is the m-dimensional real space. Given a measure $\alpha_\rho(x, y)$ on each pair of adjacent points x, y based on the values $\rho(x), \rho(y)$, we define

$$\alpha_\rho(x, y) = \begin{cases} \mu(\rho(x), \rho(y)) & \text{if } x \text{ and } y \text{ are adjacent} \\ 0 & \text{otherwise} \end{cases} \tag{12.3}$$

where $\mu : R^m \times R^m \to [0, 1]$ with $\mu(u, v) = \mu(v, u)$ and $\mu(u, u) = 1$. Note that one can define $\mu(u, u) = c$ and $\mu(u, v) \le c$ where $c \in [0, 1]$ for all u. α_ρ is used to measure "neighbor-connectivity." The next step is to develop path-connectivity so that λ-connectedness on $< G, \rho >$ can be defined in a general way.

Let $x_1, x_2, \ldots, x_n$ be a simple path in G. The path-connectivity β of a path $\pi = \pi(x_1, x_n) = \{x_1, x_2, \ldots, x_n\}$ is defined as

$$\beta_\rho(\pi(x_1, x_n)) = \min\{\alpha_\rho(x_i, x_{i+1}) | i = 1, \ldots, n-1\} \tag{12.4}$$

or

$$\beta_\rho(\pi(x_1, x_n)) = \prod\{\alpha_\rho(x_i, x_{i+1}) | i = 1, \ldots, n-1\} \tag{12.5}$$

Finally, the degree of connectedness (connectivity) between two vertices x, y with respect to ρ is defined as:

$$C_\rho(x, y) = \max\{\beta(\pi(x, y)) | \pi \text{ is a (simple) path.}\} \tag{12.6}$$

For a given $\lambda \in [0, 1]$, point $p = (x, \rho(x))$ and $q = (y, \rho(y))$ are said to be λ-connected if $C_\rho(x, y) \ge \lambda$. In image processing, $\rho(x)$ is the intensity of a point x and $p = (x, \rho(x))$ defines the pixel.

If $< G, \rho >$ is an image, then this equivalence relation can be used for segmentation or partitioning the image into different objects. On the other hand, if a potential function f is partially defined on G, then one can fit f to be ρ such that $< G, \rho >$ is λ-connected on G.

The measure μ may vary depending on the different situations. In previous work, three kinds of μ were used [5, 16]:

$$\mu_1(u,v) = \begin{cases} 1 - \frac{\|u-v\|}{\|u\|+\|v\|} & \text{if} \|u\| + \|v\| \neq 0 \\ 1 & \text{otherwise} \end{cases} \tag{12.7}$$

and

$$\mu_2(u,v) = 1 - \frac{\|u-v\|}{H}, \tag{12.8}$$

where $H = max\{\|u\| | u \in \rho(V)\}$. The following two theorems provide a sound foundation for λ-connectedness:

Theorem 12.1 ([15]). *If the connectedness function μ has the properties (i) $\mu(u,u)$ = 1, (ii) $\mu(u,v) = \mu(v,u)$, and (iii) $\mu(u,v) \geq min\{\mu(u,w),\mu(w,v)\}$, then λ-connectedness satisfies the conditions of an equivalence classification.*

Theorem 12.2 ([14]). *Let G be a simple undirected graph or an acyclic directed graph. Assume μ is given by (12.7) or (12.8). The necessary and sufficient condition under which there exists a λ-connected interpolation is that for any two vertices x and y in J, every shortest path between x and y in G is λ-connectable.*

Theorem 12.2 is an extended version of Theorem 3.1. In Theorem 12.2, if a path π is λ-connectable, then there is a valuation to each vertex a, where $a \in G - J$, on the path such that $\beta_\rho(\pi)$ is not less than λ. The philosophical meaning of λ-connectedness is that a matter or event can be split into "gradually varied" parts. In fact, the original meaning of digital continuous functions was designed for such type of purposes by Rosenfeld [32]. An object can be reconstructed based on the sample points if it is "gradually varied." We give a proof of Theorem 12.2 in Section 12.4.6.

12.3.2 λ-Connected Segmentation Algorithm

The λ-connected segmentation algorithm uses the breadth first search technique [22]. It is a fast search approach to get a connected component on a graph. First, the algorithm starts at a vertex p and checks all adjacent vertices to see if it is λ-connected to p. We then insert all λ-connected (adjacent) points (neighbors) into a queue. We "remove" a vertex from the queue and let it be p, then repeatedly do this for p's neighbors until the queue is empty. The algorithm is shown below:

Algorithm 12.1. Breadth-first-search technique for λ-connected components.
   Step 1: Let p_0 be a node in $< G, \rho >$. Set
   $\quad\quad L(p_0) \leftarrow *$ and $QUEUE \leftarrow QUEUE \cup p_0$
   $\quad\quad$i.e., labeling p_0 and sending p_0 to a queue $QUEUE$.
   Step 2: If QUEUE is empty, then proceed to Step 4; otherwise,
   $\quad\quad p_0 \leftarrow QUEUE$ (top of $QUEUE$). Then,
   $\quad\quad L(p_0) \leftarrow 0$.
   Step 3: For each p with an edge linking to p_0 , if
   $\quad\quad L(p) \neq 0, L(p) \neq *$ and $C(p,p_0) \geq \lambda$, then
   $\quad\quad QUEUE \leftarrow QUEUE \cup p$ and $L(p) \leftarrow *$. Then, go back to Step 2.
   Step 4: Stop. $S = \{p : L(p) = 0\}$ is one λ-connected part.

12.3.3 λ-Connected Segmentation for Quadtree Represented Images

This section presents a fast image segmentation algorithm based on quadtree or octree stored images. In typical image segmentation applications, the domain is a rectangular region. Quadtree and octree representations are commonly used in medical imaging and spatial databases to compress data [24, 25]. If an image is stored or compressed by a quadtree, then the algorithm presented in this section provides a method that does not require restoration or decoding of the quadtree code before the image can be used. In other words, the quadtree partition is directly used to build a graph, and then a λ-connected segmentation is performed on the new graph. The advantage of using such a strategy is to significantly increase the segmentation speed.

A compressed image represented in the quadtree shall have a leaf index with value [24, 35]. An image is split into four quadrants, namely Q_0, Q_1, Q_2 and Q_3, which represent the upper-left, upper-right, bottom-right and bottom-left quadrants, respectively. Specific formats are used to describe the structure of the compressed image in the quadtree representation. For example, $(Null, 0)$ means that the entire image is filled by "0," $(< 3 >, 128)$ means that the bottom-left quadrant is filled by "128," and $(< 2 >< 1 >, 255)$ means that the upper-right quadrant and the bottom-right quadrant of the image is filled by 255. In this example, the leaf size may be computed by: $\frac{n}{2^2}$ where n is the length of the image.

Typical image segmentation must go through each point so the time complexity must be at least $O(n^2)$ [22], where n is the length of the image and we assume $n = 2^k$. In the quadtree technique, a leaf $(< 2 >< 1 >, 255)$ will represent $n/2^2 \times n/2^2$ pixels.

Assume that the number of quadtree leaves is N. Then the segmentation algorithm can be described by first defining the adjacency graph $G_Q = (V_Q, E_Q)$ for the quadtree stored image where each leaf is a node in G_Q. If u and v are two adjacent leaves in V_Q, then $(u, v) \in E_Q$. In a 2D image, there are two types of neighborhood systems: the four-neighbor system where each point has only four neighbors and the eight-neighbor system where each point has eight neighbors.

Lemma 12.1. G_Q has at most $3N$ edges for a four-neighbor system. And G_Q has at most $4N$ edges for an eight-neighbor system.

Proof. Since each node in G_Q is a quadtree leaf or a 2^t, $t \leq k$ square in the original image, each pixel in the square has the same value (or nearly the same value due to the lossy image representation). In other words, G_Q provides a partition for the original image $< G, \rho >$. An "edge" of the partition is defined as containing at least one boundary side of a leaf. The "vertex" of the partition is at least one corner point of a leaf. Both "edges" and "vertices" are virtual to the image. However, it follows the role of Euler planar graph [22], $e = v + f - 2$, where $f = |V(G_Q)|$. The "edges" of the partition indicate the adjacency of G_Q in the four-neighbor system for 2D images. So we have $e = |E(G_Q)|$. Observing a "vertex" of the partition only connects three or four edges (except in the boundary of the image), and an "edge" connects two vertices (except in the boundary of the image). Therefore, $3v \leq 2e \leq 4v$. According to $e = v + f - 2$, we have $e = v + f - 2 \leq (2/3)e + f - 2$. If we ignore the small constant, then $e \leq 3f$. Using the same method, we have $e \geq (1/2)e + f - 2$, so $e \geq 2f$. We can prove that G_Q has at most $4N$ edges for an eight-neighbor system for 2D images. □

How much time is needed to build G_Q? It depends on how the quadtree code is stored. Basically, there are two ways. In the first way, the quadtree code is stored in the "depth-first" mode and we do the following: recursively store the first quadrant, Q_0 and its off-springs, then store Q_1 and its off-springs, and so on. In this mode, we can build G_Q quickly since we only need to compare the neighbors in the quadtree code sequence to see if they are adjacent. According to Lemma 12.1, the time complexity of the algorithm is linear. (How we store an image in quadtree format is not the focus of this book. We can recursively generate the quadtree-code for $Q_0, \ldots, Q_3$. In this method, we only store the code if it is a leaf, which would only take $O(N)$ time.) In the second way, the quadtree code is stored in the "breadth-first" mode: we sequentially store the quadtree index codes for the largest blocks, then second largest blocks, and so on. The time to get G_Q may be longer, since we need to check if the current block is adjacent to any previous block. The time complexity could be $O(N \cdot N)$.

Lemma 12.1 provides us another advantage. We only need an $O(|V_Q|)$ time algorithm to perform the segmentation using λ-connectedness. The value of $|V_Q|$ is usually much smaller than $n \cdot n$, the original image size. Even through $|V_Q|$ is dependent on the actual image, it is very reasonable to say that the average is $O(n)$. Therefore,

Theorem 12.3. *There is an $O(|V_Q|)$ time algorithm to perform segmentation using λ-connectedness.*

Without decoding the quadtree code in the original image, we cannot perform a statistical mean-based segmentation since it is not a mathematical classification. A leaf (or a block) added to a segment probably does not satisfy the requirement of $|p - mean| < \delta$ since the mean may change. We may need to break a leaf to get a more precise segmentation. Developing this idea can lead to another algorithm: (1) separate the leaf into four sub-blocks, (2) if one sub-block can merge into the segment, repeat this step, (3) insert the rest of the sub-blocks into the quadtree code sequence, then repeat. This algorithm is faster than restoring the whole image, but is slower than λ-connected quadtree segmentation. Very fast algorithms for image segmentation and analysis [23] based on cloud computing will be a significant issue in Big-Data technology in the near future [28]. We developed a λ-connected algorithm that has such an intent [19]. Today, we are at the Petabyte Age [2]. It requires fast communication speeds, huge amounts of storage, and very fast algorithms. There are many research areas that mathematicians and computer scientists can explore not only to improve technical velocity but also to build foundations to support such a change.

12.3.4 λ-Band-Connected Search Method

The λ-band-connected method proposed in this section reduces the effects of noise. The basic idea of this method is to use the average value in a neighborhood to represent the potential function value at one point. In other words, two pixels are λ-connected in a banded range if there is a λ-connected path that connects the two pixels in terms of average intensity of the band-width. This method, called λ-band-connected search, is not only for image segmentation but also for general graph based λ-connectedness.

The method can also be considered in two ways: The average band and the distribution band. In the average band, every point has equal weight, but in the distribution band, a different weight is assigned to each point.

The end pixel intensity could be the average intensity with the same band-width or different band-widths from the inner points of the linking path.

12.3.4.1 The Average Band

The band-point intensity is the average value of all intensities in the band within a circular, square, or rectangular area around a center point. Thus, two identifiable treatments can be considered. That is, the center-counted and not-center-counted. In the case where the same bandwidth is identified for the end points and the internal points, a new potential function ρ can be computed using the standard λ-connected method on $< G, \rho >$ where

$$\rho'(p) = \frac{\Sigma_{x \in N(p,B)} \rho(x)}{|N(p,B)|}, \tag{12.9}$$

and $N(p,B)$ is a circular or rectangular band/neighborhood, where B is centered on p.

The accuracy in the computation of ρ depends on the restrictions on the neighborhood and the end points. For accurate results, the end points must be noiseless or have a very low tolerance noise level.

For example, in a family relationship map, edges indicate siblings, marriages, and parent-children relations. The potential ρ is the educational or personal achievements of the individuals within the family. A member p in such a map is highly influenced by neighborhood relations. If the potential function of this member $\rho(p)$ is different in the neighborhood then such a situation is the result of an outlier or sampling noise.

12.3.4.2 The Distribution Band

The Gaussian distribution band is considered in this paper. Rather than using the average value for the potential in the distribution band, the expected value in the

Gaussian distribution is computed. It is assumed here that the band range covers only two to three standard deviations (σ), about 75–97 % of the total energy, in order to reduce the computational cost. Therefore, the potential of a member p is computed as:

$$\rho'(p) = \int_{x \in N(p,B)} e^{\frac{-d^2(x,p)}{\sigma^2}} \rho(x) \qquad (12.10)$$

where $d(x,p)$ is the distance between x and p and may be non-Euclidean distance. As previously discussed, connected components can be computed from new and original potential functions, where bandwidths are the same. For situations where bandwidths are not the same, it is necessary to work on both new and original potential functions, although the transition property to preserve an equivalence relation might not exist. In most cases, the partition using a small bandwidth is the refinement of using a bigger bandwidth.

As we discussed above, if two bandwidths are the same, we could use a new potential function and the original G to get the connected components. In this way, we work on the averaged "image." If the two bandwidths are not the same, we have to work on both the new and the old potential functions. In addition, it does not necessarily require the transitive property to preserve an equivalence relation. This is because a segmentation can be done by a similar relation that only holds reflexive and symmetric laws.

12.3.4.3 The Boundary Band

For some applications, the result is certain when a pixel value is in a range. It may not be certain when the pixel value is at the boundary of the range. Even though, we can still use the methods described above, it wastes time when the pixel value is in the range and the result is certain. Here, we propose a boundary band that only does extra and careful calculations when the pixel value is definitely on the boundary band. This philosophy comes from rough sets [31].

In this algorithm, we split the pixels into three categories: $inRange = 1$, $outRange = 0$, and $boundaryBand = fuzzyvalue$. We only do a λ-connected search or a λ-band-connected search for the boundary band category.

12.4 Expanding Gradually Varied Functions to Intelligent Data Reconstruction

Generally, a λ-connected fitting can be described as follows: We are given an undirected graph $G = (V,E)$ or a directed graph $G = (V,A)$ and a subset J of V. If $\rho_J : J \to R^m$ is known, then the λ-connected fitting finds an extension/approximation

ρ of ρ_J such that $\rho : V \to R^m$ is λ-connected on G (meaning that $< G, \rho >$ has only one λ-connected component) for a certain λ.

With the concept of λ-connected fitting defined, the first question to ask is whether or not a λ-connected fitting exists and how to determine the λ-connected fitting function.

Chen, Cooley, and Zhang described an intelligent data fitting method to reconstruct a velocity volume in [17]. The key idea is to perform a λ-connected segmentation first, then fit the data. An example of this approach is presented in Chap. 8.

In this chapter we mainly discuss the fitting using derivatives [14]. We provided the primary fitting theorem for λ-connectedness, Theorem 12.2, in Sect. 12.3.

12.4.1 Three Mathematical Issues of λ-Connected Fitting

There are some mathematical issues regarding λ-connected fitting related to the following: (1) λ-connected approximation when there is no λ-connected interpolation with respect to guiding points, (2) optimum fitting, and (3) λ value determination.

First, the λ-connected interpolation or gradually varied interpolation may not always exist. An approximation method can be used to get a λ-connected fitting in terms of optimal uniform approximation [10]. For a least squares fitting, since it relates to derivatives, more tools and knowledge are required to deal with this problem.

Second, the λ-connected fitting may not be unique and in such a case, how do we choose the best fitting solution? Basically, the answer depends on the different criteria. More research on this issue will be done in the near future.

An alternative solution can be found by performing a λ-connected fitting first because it can handle the irregular domain easily. Then, use a standard fitting method for the secondary process. For instance, we can first generate a gradually varied surface, then use the B-spline method to do a fitting if a smooth surface is required.

Third, the calculation of a value for λ in either segmentation or fitting is a critical issue. For segmentation, the selection of λ depends on how many segments we want to separate. In the following section, we discuss the maximum-connectivity spanning tree method, which can be used to find the value of λ. Since the process of finding the maximum-connectivity spanning tree is time consuming, we may use a histogram or an experimental test to solve this problem.

A binary search method can be used to test the value of λ. First, test for $\lambda = 0.5$. If λ is too large, then test for $\lambda = 0.25$. If the λ is too small, then test for $\lambda = 0.75$ and so on. Similarly, the same method may be used for λ-connected fitting.

12.4.2 λ-Connected Fitting with Gradients

The above λ-connected fitting methods only deal with "continuous" surfaces or functions. This section will present a new method for surface fitting with smooth derivatives (we call it the differentiable λ-fitting method). It will allow negative values for the gradients.

As we know, the purpose of λ-connected fitting is to fit a surface without a mathematical formula assumption or we do not know the formula and do not want to make an assumption. In addition, the domain may be irregular so the B-spline or other classical fitting methods cannot be applied appropriately.

There are three problems related to differentiable surface fitting using λ-connectedness that range from simple to complex.

Case 1: Given $f_J : J \to R, J \subset D$ and $\frac{df}{dv} : D \to R$, and we want to get $f = f_D$.

Case 2: Given $f_J : J \to R, J \subset D$ and $\frac{df}{dv} : J' \to R, J' \subset D$, and we want to get f_D.

Case 3: Given $f_J : J \to R, J \subset D$ and $\frac{df}{dv} : D \to R, \ldots \ldots, \frac{df^k}{dv^k} : D \to R$, and we want to get f_D.

We present an approach for each case in the following sections.

12.4.3 Case 1: With Complete First-Order Gradients

It is important that we assume that there is a way to calculate gradients using $< D, f >$. In addition, we need an iteration process so we can determine what is the most fitted "fitting."

If D is two dimensional (it does not necessarily have to be a 2D domain), we know $f_J : J \to R$ and $\frac{df}{dv} : D \to R$. Then, the idea of an algorithm can be developed as follows. For a specific problem, we know how to calculate gradients based on the discrete values on D. For example, for a rectangular area, we can use the difference method to get the derivatives.

Since at the beginning of the fitting, only some of the f values are available, the gradient calculation is not finalized. An iteration process is designed for the approximation. It is easy to see if the domain has its triangulation, we can perform a similar calculation.

Algorithm 12.2. λ-connected fitting with complete first-order gradients.

Step 1: Check if f_J satisfies the λ-connected interpolation condition. We can assume that $\frac{df}{dv}$ is λ-connected. Start at a point p in J, get a point q from its neighborhood N_p. ($N_p = N(p, B)$, where B contains all adjacent vertices of p.)

Step 2: Obtain all $f(q)$ such that $f_{J \cup \{q\}}$ satisfies the λ-connected interpolation condition. Put all of $f(q)$'s into a set $F(q)$.

Step 3: Select a f_0 in $F(q)$ such that the gradient on q (which is only based on the existing set $f_{J \cup \{q\}}$) is the closest to $\frac{df}{dv}$ at point q. According to Theorem 12.2,

$F(q)$ can be determined. For instance, we can use the difference analysis method to get the gradient from the values of the points [14]. Search for the best combination from all possible combinations for q. Select a new neighbor of p. Repeat from Step 2 until all neighbors are fitted.

Step 4: Select the new guiding point p sequentially or randomly. Select a neighbor of p and go to Step 2, repeats until all points are fitted in D.

Step 5: For all points that have new values, calculate the gradients. It should be different from the first time the gradients were calculated since all values of f are now known. Find the best $f(p) \in F(p)$ such that the "actual" $\frac{df}{dv}$ (based on f) is the closest to the "ideal" gradient $\frac{df}{dv}$ known before the interpolation. f must remain λ-connected during the entire process.

12.4.4 Case 2: With Incomplete First-Order Gradients

There are two ways to solve the second problem. First, we fit λ-connected gradients $\frac{df}{dv}$ on D based on $\frac{df}{dv}$ in J'. Then we can use the algorithm in Case 1 to get the final solution.

Secondly, if we do not know the gradient values at all points in D, it is still possible to use the existing ones to determine the best fit. We just need to check the gradient values for the points that are in J'. The rest of the process is the same as the process described in Case 1.

12.4.5 Case 3: With High Order Gradients

It is not necessary to restrict a domain to a rectangle in order to calculate higher degrees of derivatives and gradients. When we need to get a surface with a higher degree of smoothness, for example when $\frac{d^2 f}{dv^2}$ is known, we can define a set that is the neighborhood of a neighborhood of the point p to decide the fitting in the original graph/domain. This design philosophy was used in the finite elements method.

For Case 3, suppose we know how to calculate $\frac{d^i f}{dv^i}$ based on the values of f, and we have all the values of $\frac{df}{dv}: D \to R, \ldots \ldots, \frac{d^k f}{dv^k}: D \to R$.

We can use an algorithm similar to that of Case 1 to calculate f from f_J, to best fit the known values : $\frac{d^i f}{dv^i}, i = 1, \ldots, k$. If a conflict occurs, we would give more weight to the lower order $\frac{d^i f}{dv^i}$. An iteration process may apply in this situation. In fact, we only need to change Steps 3 and 5, the rest of the steps would remain the same.

Step 3′: Select a f_0 in $F(q)$ such that the gradient on q (only based on the existing set $f_{J\cup\{q\}}$) is the closest to $\frac{d^i f}{dv^i}$, where $i = 1, \ldots, k$ at point q. According to Theorem 2.2, $F(q)$ can be determined. For instance, we can use the difference analysis method to get the gradient from the values of points. We then search for the best combination from all possible combinations for q. Selecting a new neighbor of p, go to Step 2, and repeat the process until all neighbors are fitted.

Step 5′:   For all points that have new values, calculate the new gradients. These new gradients would be different from the first time the values were determined since now all values of f are known. Find the best $f(p) \in F(p)$ such that the "actual" $\frac{d^i f}{dv^i}$, $i = 1,\ldots,k$ (based on f), is closest to the "ideal" gradient $\frac{d^i f}{dv^i}$, $i = 1,\ldots,k$, which was known before the interpolation. This process must maintain f as λ-connected at all times.

12.4.6 Proof of the Theorem for λ Connected Fitting

A λ-connectable path means that there are values for all points on the path such that the path is λ-connected. Two pixels p and q (vertices with their potential function values) are λ-connectable if there is a λ-connectable path linking these two points.

Two pixels p and q are said to be normal λ-connectable if every path linking these two points is λ-connectable.

$< G, f_J >$ is normal λ-connectable if every pair of points is normal λ-connectable. Let S be a subgraph of G. $< S, f_{J \cap S}$ is normal λ-connectable if every pair of points in $< S, f_{J \cap S} >$ is normal λ-connectable.

A λ-connected interpolation provides a valuation for f_G such that every path in G is λ-connected. We have,

Proposition A   The necessary condition of the existence of a λ-connected interpolation is that $< G, f_J >$ is normal λ-connectable.

We want to prove that the condition in Proposition A is also the sufficient condition shown below.

Theorem 12.2   Let G be a simple undirected graph or an acyclic directed graph. Assume that μ is given by (12.7) or (12.8). The necessary and sufficient condition under which there exists a λ-connected interpolation is that for any two vertices x and y in J, every shortest path between x and y in G is λ-connectable.

Before we prove this theorem, we present the two lemmas [9].

Lemma A   Let p=(x,f(x)) and q = (y,f(y)). If path $P = p_0 \ldots p_k$ is λ-connectable, then

(1) When (12.7) applies, assume $f(x) \geq f(y)$, $f(x) \neq= 0$, and $f(y) \neq= 0$, then $\frac{f(x)}{f(y)} \leq (\frac{2-\lambda}{\lambda})^k$, or
(2) When (12.8) applies, assume $f(x) \geq f(y)$, so $f(x) - f(y) \leq k(1-\lambda)M$.

and vice versa.

Lemma B   For any two points p=(x,f(x)) and q = (y,f(y)) in $< G, f_J >$, if the distance between x and y is k, then

(1) When (12.7) applies, $f(x) \geq f(y)$, $f(x) \neq = 0$, and $f(y) \neq = 0$, so $\frac{f(x)}{f(y)} \leq (\frac{2-\lambda}{\lambda})^k$,
   or
(2) When (12.7) applies, $f(x) \geq f(y)$, so $f(x) - f(y) \leq k(1 - \lambda)M,.$

Therefore, $< G, f_J >$ is normal λ-connectable.

Proof of Theorem 12.2 The proving strategy is similar to the proof of the theorem in [6]. We only give the proof when (12.7) applies. If (12.8) applies, a very similar method is used, as described in [6]. Let $x \in J$ have an adjacent neighbor $y \notin J$. We want to determine the value $f(y)$ so that $< G, f_{J'} >$, where $J' = J \cup \{y\}$, is normal λ-connectable. If Eq. (12.8) applies, set $f(y) = f(x)$. If there is a vertex $x' \in J$, such that $(y, f(y))$ and $(x', f(x'))$ are not normal λ-connectable, then $f(y)/f(x') > ((2-\lambda)/\lambda)^{d(y,x')}$ (we might as well assume $f(y) > f(x')$).

Since x, y are adjacent, $d(x, x') \leq d(y, x') + 1$. Because $< G, f_J >$ is normal λ-connectable, $f(x)/f(x') < ((2-\lambda)/\lambda)^k$, $d(x, x') = k$. So, $k = d(y, x') + 1$. We have

$$((2-\lambda)/\lambda)^{k-1} < f(x)/f(x') \leq ((2-\lambda)/\lambda)^k.$$

Let $f(y) = f(x)/(2-\lambda)/\lambda$.

Because of the new $f(y)$, $f(y)/f(x') = (f(x)/(2-\lambda)/\lambda))/f(x') \leq ((2-\lambda)/\lambda)^{k-1}$. That is to say, every path between y, x' is normal λ-connectable since every such path has a length larger than $k - 1$.

Is there any other point x'' in J that can break this "new deal"? If such a point cannot be found, then we have the final valuation $f(y)$., then

$f(y)/f(x'') > (2-\lambda)/\lambda))^{d(y,x'')}$ or
$f(x'')/f(y) > (2-\lambda)/\lambda))^{d(y,x'')}.$

We can easily see that if $f(x) > f(x'')$, then the above inequalities will not be possible. We can also see that if $d(y, x'') = d(x, x'') + 1$, then the above inequalities will also be impossible.

We only need to show the case where:

(1) $f(x'') > f(x)$, and
(2) $d(y, x'') = d(x, x'')$, or $d(y, x'') = d(x, x'') - 1$.

We already know that $d(x, x') = k$, $d(y, x') = k - 1$, and $f(y)/f(x') > ((2-\lambda)/\lambda)^{k-1}$.

We derive a contradiction if $f(x'')/f(y) > (2-\lambda)/\lambda))^{d(y,x'')}$. We know $f(x'')/f(x') = f(x'')/f(y) \times f(y)/f(x')$. Therefore, $f(x'')/f(x') = (2-\lambda)/\lambda))^{d(y,x'')} \times ((2-\lambda)/\lambda)^{k-1}$, so

$$f(x'')/f(x') > (2-\lambda)/\lambda))^{d(y,x'')+k-1}. \tag{12.11}$$

On the other hand, $d(x'', x')$ indicates the length of the shortest path between x' and x''. Thus, $d(x'', x') \leq d(x'', y) + d(y, x')$, i.e. $d(x'', x') \leq d(y, x'') + k - 1$.

Because $x', x'' \in J$, we have $f(x'')/f(x') \leq (2-\lambda)/\lambda))^{d(x',x'')} \leq (2-\lambda)/\lambda))^{d(y,x'')+k-1}$. This contradicts (12.11).

Corollary A If both $< q, f'(q) >$ and $< q, f''(q) >$ can make $< G, f_{J \cup \{q\}} >$ normal λ-connectable, then any value $g(q)$, where $f'(q) \leq g(q) \leq f''(q)$, would make $< G, f_{J \cup \{q\}} >$ normal λ-connectable.

Acknowledgements Many thanks to Dr. Peter Schroeder at Caltech for the online software on Chaikin's method and the four point algorithm:
http://www.multires.caltech.edu/teaching/demos/java/chaikin.htm.
http://www.multires.caltech.edu/teaching/demos/java/4point.htm.
Also thanks to Dr. Ken Joy at UC Davis for pointing out Chaikin's work. Many thanks to Dr. Paul Chew at Cornell University for his software on Delaunay diagrams. http://www.cs.cornell.edu/home/chew/Delaunay.html.

References

1. Belytschko T, Krongauz Y, Organ D, Fleming M, Krysl P (1996) Meshless methods: an overview and recent developments. Comput Methods Appl Mech Eng 139(1–4):3–47
2. Brumfiel G (2011) Down the petabyte highway, Nature 469:282–283
3. Catmull E, Clark J (1978) Recursively generated B-spline surfaces on arbitrary topological meshes. Comput Aided Des 10(6):350–355
4. Chaikin G (1974) An algorithm for high speed curve generation. Comput Graph Image Process 3:346–349
5. Chen L (1985) Three-dimensional fuzzy digital topology and its applications(I). Geophys Prospect Pet 24(2):86–89
6. Chen L (1990) The necessary and sufficient condition and the efficient algorithms for gradually varied fill. Chin Sci Bull 35:870–873 (Its Chinese version was published in 1989.)
7. Chen L (1990) Gradually varied surfaces and gradually varied functions, in Chinese, 1990; in English 2005 CITR-TR 156, University of Auckland
8. Chen L (1991) The lambda-connected segmentation and the optimal algorithm for split-and-merge segmentation. Chin J Comput 14:321–331
9. Chen L (1992) Random gradually varied surface fitting. Chin Sci Bull 37(16):1325–1329
10. Chen L (1994) Gradually varied surface and its optimal uniform approximation. In: *IS&T SPIE* symposium on electronic imaging, SPIE Proceeding, San Jose, vol 2182, pp 300–307
11. Chen L (2004) Discrete surfaces and manifolds. Scientific and Practical Computing, Rockville
12. Chen L (2010) A digital-discrete method for smooth-continuous data reconstruction. Capital Science 2010 of The Washington Academy of Sciences and its Affiliates, 27–28 Mar 2010
13. Chen L (2010) Digital-discrete surface reconstruction: a true universal and nonlinear method. http://arxiv.org/ftp/arxiv/papers/1003/1003.2242.pdf
14. Chen L, Adjei O (2004) Lambda-connected segmentation and fitting. In: Proceedings of IEEE international conference on systems man and cybernetics, vol 4. IEEE, Hague, pp 3500–3506
15. Chen L, Berkey FT, Johnson SA (1994) The application of a fuzzy object search technique to geophysical data processing. In: Proceedings of *IS&T SPIE* symposium on electronic imaging, SPIE Proceeding, vol. 2180. SPIE, San Jose, pp 300–309
16. Chen L, Cheng HD, Zhang J (1994) Fuzzy subfiber and its application to seismic lithology classification. Inf Sci 1(2):77–95
17. Chen L, Cooley DH, Zhang L (1998) An Intelligent data fitting technique for 3D velocity reconstruction. In: Application and science of computational intelligence, proceeding SPIE 3390. SPIE, Orlando, pp 103–112
18. Chen L, Adjei O, Cooley DH (2000) λ-connectedness: method and application. In: Proceedings of IEEE conference on system, man, and cybernetics 2000. IEEE, Nashville, pp 1157–1562

19. Chen L, Zhu H, Cui W (2006) Very fast region-connected segmentation for spatial data. In: Proceedings of IEEE international conference on systems, man and cybernetics, Hong Kong, pp 4001–4005
20. Cleveland WS, William S (1979) Robust locally weighted regression and smoothing scatterplots. J Am Stat Assoc 74(368):829–836
21. Doo D, Sabin M (1978) Behaviour of recursive division surfaces near extraordinary points. Comput Aided Des 10:356–360
22. Cormen TH, Leiserson CE, Rivest RL (1993) Introduction to algorithms. MIT, Cambridge
23. Fisher R , Ken D, Fitzgibbon A, Robertson C, Trucco E (2005) Dictionary of computer vision and image processing, Wiley, Hoboken
24. Gonzalez RC, Wood R (1993) Digital image processing. Addison-Wesley, Reading
25. Hearn D, Baker MP (2004) Computer graphics (with OpenGL), 3rd edn. Prentice Hall, Upper Saddle River
26. Lancaster P, Salkauskas K (1981) Surfaces generated by moving least squares methods. Math Comput 87:141–158
27. Levin D (2003) Mesh-independent surface interpolation. In: Brunnett G, Hamann B, Mueller K, Linsen L (eds) Geometric modeling for scientific visualization. Springer, Berlin/London
28. Lynch C (2008) Big data: How do your data grow? Nature 455(7209):28–29
29. Mount DM (2002) Computational geometry, UMD lecture notes CMSC 754. http://www.cs.umd.edu/~mount/754/Lects/754lects.pdf
30. Pavilidis T (1982) Algorithms for graphics and image processing. Computer Science Press, Rockville
31. Pawlak Z (1999) Rough sets, rough functions and rough calculus. In: Pal SK, Skowron A (eds) Rough fuzzy hybridization. Springer, New York, pp 99–109
32. Rosenfeld A (1986) 'Continuous' functions on digital pictures. Pattern Recognit Lett 4:177–184
33. Rosenfeld A, Kak AC (1982) Digital picture processing, 2nd edn. Academic, New York
34. Russell S, Norvig P (2003) Artificial intelligence: a modern approach, 2nd edn. Prentice Hall, Englewood Cliffs
35. Theodoridis S, Koutroumbas K (2003) Pattern recognition, 2nd edn. Academic, San Diego

Glossary

Algorithm A set of steps of instructions for solving a problem.

Digital space A grid space in which every element is a integer point.

Discrete space A graph with geometric distance measure or metric.

Digital image A digital function from a 2D domain to integers.

Sample Points A group of samples collected from a region that is usually a rectangle, but could be any other shapes. The sample point has two components: one is the location as (x,y) in 2D or (x,y,z) in 3D; another is the value in real numbers.

Domain The mathematical term for the region that holds locations. Range or co-domain is usually used as the values.

Interpolation Extend the sample data (usually a few points) into entire region or domain. It will be the function (or surface for 2D) for the region. Interpolation requires the new function passes the sampling values at each sample points.

Approximation Similar to interpolation, but approximation does not require the new function passes the sampling values at each sample points. Because of that, approximation is usually easy to get smoothness.

Fitting Means either interpolation or approximation.

Boundary of Domain The Edge of domain.

Digital function and discrete function Domain of the function is a digital space. A 2D digital space usually means a 2D grid points in 2D Euclidean space. Discrete function is the function that is defined on discrete space.

Gradually varied function Gradually varied function is a type of discrete functions where the values of each pair of neighbors are the sample or only have small change. Gradually varied functions use A1, Am to represent the value and level changes. It is essentially to translate the discrete function into digital functions in terms of algebra. This method has the limitation to deal with derivatives.

L.M. Chen, *Digital Functions and Data Reconstruction: Digital-Discrete Methods,*
DOI 10.1007/978-1-4614-5638-4, © Springer Science+Business Media, LLC 2013

Digital-discrete reconstruction method Reinstall the discrete value Ai to the fit-ted surface. Calculate the derivative based on the discrete function not digital levels. Then, use put it into the digital levels to at the sample points to do the gradually varied fitting for the derivatives. This type of alternations backing-forward in digital and discrete functions are called Digital-discrete reconstruction method.

Taylor Expansion Any analytic function has its expansion using the Taylor series. Use the Taylor series we can get a smooth functions in certain degree.

Index